An Introduction to the
Optical Spectroscopy
of Inorganic Solids

An Introduction to the Optical Spectroscopy of Inorganic Solids

J. García Solé, L.E. Bausá and D. Jaque
Universidad Autónoma de Madrid, Madrid, Spain

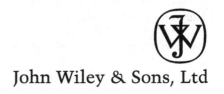

John Wiley & Sons, Ltd

Copyright © 2005   John Wiley & Sons Ltd, The Atrium, Southern Gate, Chichester,
                        West Sussex PO19 8SQ, England

                        Telephone   (+44) 1243 779777

Email (for orders and customer service enquiries): cs-books@wiley.co.uk
Visit our Home Page on www.wileyeurope.com or www.wiley.com

Other Wiley Editorial Offices

John Wiley & Sons Inc., 111 River Street, Hoboken, NJ 07030, USA

Jossey-Bass, 989 Market Street, San Francisco, CA 94103-1741, USA

Wiley-VCH Verlag GmbH, Boschstr. 12, D-69469 Weinheim, Germany

John Wiley & Sons Australia Ltd, 33 Park Road, Milton, Queensland 4064, Australia

John Wiley & Sons (Asia) Pte Ltd, 2 Clementi Loop #02-01, Jin Xing Distripark, Singapore 129809

John Wiley & Sons Canada Ltd, 22 Worcester Road, Etobicoke, Ontario, Canada M9W 1L1

Wiley also publishes its books in a variety of electronic formats. Some content that appears
in print may not be available in electronic books.

Library of Congress Cataloging-in-Publication Data

García Solé, J. (José)
    An introduction to the optical spectroscopy of inorganic solids / J. García Solé,
and L. E. Bausá, and D. Jaque.
      p. cm.
    Includes index.
    ISBN 0-470-86885-6 (cloth)—ISBN 0-470-86886-4 (pbk.)
    1. Solids—Spectra.   2. Energy-band theory of solids.   3. Solid state chemistry.
4. Chemistry, Inorganic.   5. Spectrum analysis.
    I. Bausá, L. E. (Louisa E.)   II. Jaque, D. (Daniel)   III. Title.

    QC176 .8 . O6G25   2005
    530 .4′1—dc22   2004018408

British Library Cataloguing in Publication Data

A catalogue record for this book is available from the British Library

ISBN 0-470-86885 6 (cloth)
    0-470-86886 4 (paper)

Typeset in 10/12pt Times by TechBooks, New Delhi, India

To my wife, Rosario, and my two children, Pepe and Pablo. They are the most important part of my life.

José García Solé

To Beatriz, Carmen, Fernando and Luis for their love.

Luisa Bausá López

To my family

Daniel Jaque Garcia

Contents

Preface

This book treats the most basic aspects to be initiated into the field of the optical spectroscopy of solids, so that a student with some background in quantum physics, optics, and solid state physics may be able to interpret simple optical spectra (absorption, reflectivity, emission, scattering, etc.) and learn about the main basic instrumentation used in this field.

The term 'optical spectroscopy' refers only to the range of interacting electromagnetic radiation lying within the so-called 'optical range'; a range that includes the visible and a small part of the ultraviolet and infrared spectral regions, at about 200–3000 nm. We improperly label this radiation as 'light', while this term only strictly refers to that radiation which can be detected by the human eye. The term 'solids' includes metals, semiconductors and insulators. Although most of the material treated in this book concerns the spectroscopy of centers embedded in inorganic materials, the principles described here are also applicable to molecules and atoms in the gaseous and/or the liquid state.

Although a number of excellent books covering the field of optical spectroscopy are available, they are mostly extensive books, due to their systematic and formal contents. Thus, we aimed to write this book for a number of specific reasons:

(i) Members of several scientific communities (analytical chemistry, solid state physics, photonics, etc.) may be interested in a simple introductory book, since the basic concepts of spectroscopy and the instrumentation treated in this text apply to solid as well as to molecular systems.

(ii) An introductory book is appropriate because a number of optical spectroscopic techniques are used in many laboratories for material characterization.

(iii) Spectroscopy is a topic that is now included in several courses for undergraduate and postgraduate students.

(iv) The research area of optical materials is, at present, an activity within the modern and more general area of photonics. A great variety of optical materials are based on inorganic materials activated with optically active ions (centers).

Any experiment involving optical spectroscopy consists of a light source, a sample, and a detection–recording system. According to this scheme, this book is organized as follows.

The book starts with a short introduction to the fundamentals of optical spectroscopy, (Chapter 1) describing the basic standard equipment needed to measure optical spectra and the main optical magnitudes (the absorption coefficient, transmittance, reflectance, and luminescence efficiency) that can be measured with this equipment. The next two chapters (Chapters 2 and 3) are devoted to the main characteristics and the basic working principles of the general instrumentation used in optical spectroscopy. These include the light sources (lamp and lasers) used to excite the crystals, as well as the instrumentation used to detect and analyze the reflected, transmitted, scattered, or emitted light.

Chapter 4, presents details of the absorption and reflectivity spectra of pure crystals. The first part of this chapter connects the optical magnitudes that can be measured by spectrophotometers with the dielectric constant. We then consider how the 'valence electrons' of the solid units (atoms or ions) respond to the electromagnetic field of the optical radiation. This establishes a frequency dependence of the dielectric constant, so that the absorption and reflectivity spectrum (the transparency) of a solid can be predicted. The last part of this chapter focuses on the main features of the spectra associated with metals, insulators, and semiconductors. The absorption edge and excitonic structure of band gap (semiconductors or insulator) materials are also treated.

Chapters 5 and 6 deal with the spectra of 'optically active centers.' The term 'optically active center' corresponds to a dopant ion and its environment (or to a color center), which produces absorption and/or emission bands that are different to those of the pure crystalline host. This is the case for a large variety of optical materials, such as phosphors, solid state lasers, and amplifiers.

In Chapter 5, we discuss in a simple way static (crystalline field) and dynamic (coordinate configuration model) effects on the optically active centers and how they affect their spectra (the peak position, and the shape and intensity of optical bands). We also introduce nonradiative depopulation mechanisms (multiphonon emission and energy transfer) in order to understand the ability of a particular center to emit light; in other words, the competition between the mechanisms of radiative de-excitation and nonradiative de-excitation.

Chapter 6 is devoted to discussing the main optical properties of transition metal ions ($3d^n$ outer electronic configuration), trivalent rare earth ions ($4f^n5s^25p^6$ outer electronic configuration), and color centers, based on the concepts introduced in Chapter 5. These are the usual centers in solid state lasers and in various phosphors. In addition, these centers are very interesting from a didactic viewpoint. We introduce the Tanabe–Sugano and Dieke diagrams and their application to the interpretation of the main spectral features of transition metal ion and trivalent rare earth ion spectra, respectively. Color centers are also introduced in this chapter, special attention being devoted to the spectra of the simplest F centers in alkali halides.

Chapter 7 is a very simple introduction to 'group theory' and its usefulness to interpreting the optical spectra of active centers. The purpose of this chapter is to present some basic concepts, for non-specialists in group theory, so they can evaluate its potential and, hopefully the feasibility of applying it to simple problems, such as the determination and labeling of the energy levels of an active center by means of the character table of its symmetry group.

Finally, the book includes a collection of illustrative examples and a variety of specifically selected spectra. A number of these spectra correspond to systems that have actually been investigated in our laboratory.

José García Solé
Luisa Bausá
Daniel Jaque
Madrid, June 2004

Acknowledgments

This book emerged as a result of the experience accumulated by all of us during several years of teaching and research activity in the field of optical spectroscopy of solids. In particular, one of us (José García Solé) has taught 'Optical Solid State Spectroscopy' for several years, as a fourth-month course during the last year of the physics degree course at the Autonomous University of Madrid. We are indebted to our colleagues (Professors José Manuel Calleja and Fernando Cussó), who taught and researched this topic for several years in our university, and who provided several additional ideas for this book. Finally, we would like to thank Professors Francisco Jaque and Julio Gonzalo for critically reading of some parts of this book and, in particular, Dr Juan José Romero for helping us with a patient and critical analysis of all the chapters in the book, including both the text and the figures.

Some Physical Constants of Interest in Spectroscopy

Constant	Symbol	MKS value	Other usual units
Electron mass	m_e	9.11×10^{-31} kg	9.11×10^{-28} g
Proton mass	m_p	1.67×10^{-27} kg	1.67×10^{-24} g
Electron charge	e	1.60×10^{-19} C	4.80×10^{-10} stc
Speed of light in vacuum	c	3×10^8 m s^{-1}	3×10^{10} cm s^{-1}
Planck's constant	h	6.63×10^{-34} J s	4.14×10^{-15} eV s
	$\hbar$	1.05×10^{-34} J s	6.58×10^{-16} eV s
Permittivity of vacuum	ε_0	8.85×10^{-12} N^{-1} m^{-2} C^2	
Bohr radius	a_0	0.53×10^{-10} m	0.53×10^{-8} cm
Avogadro's constant	N_A	6.02×10^{23} mol^{-1}	
Boltzmann's constant	K	1.38×10^{-23} J K^{-1}	8.62×10^{-5} eV K^{-1}
Stefan–Boltzmann constant	σ	5.67×10^{-8} W m^{-2} K^{-4}	

A Periodic Table of the Elements for Optical Spectroscopy

Group I	Group II											Group III	Group IV	Group V	Group VI	Group VII	Group 0
H^1 +1, −1 1.008 $1s^1$																	He^2 0 4.00 $1s^2$
Li^3 +1 6.94 $2s^1$	Be^4 +2 9.01 $2s^2$											B^5 +3 10.81 $2s^2 2p^1$	C^6 +2, +4, −4 12.01 $2s^2 2p^2$	N^7 +1, +2, ±3, +4, +5 14.01 $2s^2 2p^3$	O^8 −2 15.99 $2s^2 2p^4$	F^9 −1 18.99 $2s^2 2p^5$	Ne^{10} 0 20.18 $2s^2 2p^6$
Na^{11} +1 22.99 $3s^1$	Mg^{12} +2 24.31 $3s^2$											Al^{13} +3 26.98 $3s^2 3p^1$	Si^{14} +4, −4 28.09 $3s^2 3p^2$	P^{15} +3, +5, −3 30.97 $3s^2 3p^3$	S^{16} +4, +6, −2 32.07 $3s^2 3p^4$	Cl^{17} ±1, +3, +5, +7 35.45 $3s^2 3p^5$	Ar^{18} 0 39.95 $3s^2 3p^6$
K^{19} +1 39.09 $4s^1$	Ca^{20} +2 40.08 $4s^2$	Sc^{21} +3 44.96 $3d^1 4s^2$	Ti^{22} +2, +3, +4 47.87 $3d^2 4s^2$	V^{23} +2 to +5 50.94 $3d^3 4s^2$	Cr^{24} +2, +3, +4 +6 52.00 $3d^5 4s^1$	Mn^{25} +2 to +7 54.94 $3d^5 4s^2$	Fe^{26} +2, +3 55.84 $3d^6 4s^2$	Co^{27} +2, +3 58.93 $3d^7 4s^2$	Ni^{28} +2, +3 58.69 $3d^8 4s^2$	Cu^{29} +1, +2 63.55 $3d^{10} 4s^1$	Zn^{30} +2 65.41 $3d^{10} 4s2$	Ga^{31} +3 69.72 $4s^2 4p^1$	Ge^{32} +2, +4 72.64 $4s^2 4p^2$	As^{33} +3, +5, −3 74.92 $4s^2 4p^3$	Se^{34} +1, +4, +6, ±2 78.96 $4s^2 4p^4$	Br^{35} +1, +3, +5, −1 79.90 $4s^2 4p^5$	Kr^{36} 0 83.80 $4s^2 4p^6$

Period 5 (Rb–Xe)

Property	Rb^{37}	Sr^{38}	Y^{39}	Zr^{40}	Nb^{41}	Mo^{42}	Tc^{43}	Ru^{44}	Rh^{45}	Pd^{46}	Ag^{47}	Cd^{48}	In^{49}	Sn^{50}	Sb^{51}	Te^{52}	I^{53}	Xe^{54}
Valences	+1	+2	+3	+4	+3, +5	+2 to +6	+4, +6, +7	+2 to +6	+3	+2, +4	+1	+2	+3	+2, ±4	+3, +5, −3	+4, +6, −2	±1, +3, +5, +7	0, +2, +4, +6
Atomic Weight	85.47	87.62	88.91	91.22	92.91	95.94	(98)	101.07	102.91	106.42	107.87	112.41	114.82	118.71	121.76	127.60	126.90	131.29
Config	$5s^1$	$5s^2$	$4d^1 5s^2$	$4d^2 5s^2$	$4d^4 5s^1$	$4d^5 5s^1$	$4d^6 5s^1$	$4d^7 5s^1$	$4d^8 5s^1$	$4d^{10}$	$4d^{10} 5s^1$	$4d^{10} 5s^2$	$5s^2 5p^1$	$5s^2 5p^2$	$5s^2 5p^3$	$5s^2 5p^4$	$5s^2 5p^5$	$5s^2 5p^6$

Period 6 (Cs–Rn)

Property	Cs^{55}	Ba^{56}	La^{57}	Hf^{72}	Ta^{73}	W^{74}	Re^{75}	Os^{76}	Ir^{77}	Pt^{78}	Au^{79}	Hg^{80}	Tl^{81}	Pb^{82}	Bi^{83}	Po^{84}	At^{85}	Rn^{86}
Valences	+1	+2	+2, +3	+2, +3, +4	+5	+4, +5, +6	+3 to +7	+1 to +8	+3, +4, +5	+2, +4	+1, +3	+1, +2	+1, +3	+2, +4	+3, +5	±2, +4	±1	0
Atomic Weight	132.91	137.33	138.91	178.49	180.95	183.84	186.21	190.23	192.22	195.08	196.97	200.59	204.38	207.2	208.98	(209)	(210)	(222)
Config	$6s^1$	$6s^2$	$5d^1 6s^2$	$4f^{14} 5d^2 6s^2$	$5d^3 6s^2$	$5d^4 6s^2$	$5d^5 6s^2$	$5d^6 6s^2$	$5d^7 6s^2$	$5d^9 6s^1$	$5d^{10} 6s^1$	$5d^{10} 6s^1$	$6s^2 6p^1$	$6s^2 6p^2$	$6s^2 6p^3$	$6s^2 6p^4$	$6s^2 6p^5$	$6s^2 6p^6$

Period 7 (Fr–Rf)

Property	Fr^{87}	Ra^{88}	Ac^{89}	Rf^{104}
Valences	+1	+2	+3	
Atomic Weight	(223)	(226)	(227)	(261)
Config	$7s^1$	$7s^2$	$6d^1 7s^2$	$6d^2 7s^2$

Lanthanides

Property	Ce^{58}	Pr^{59}	Nd^{60}	Pm^{61}	Sm^{62}	Eu^{63}	Gd^{64}	Tb^{65}	Dy^{66}	Ho^{67}	Er^{68}	Tm^{69}	Yb^{70}	Lu^{71}
Valences	+3, +4	+3	+3	+3	+2, +3	+2, +3	+3	+3	+3	+3	+3	+3	+2, +3	+3
Atomic Weight	140.12	140.91	144.24	(145)	150.36	151.96	157.25	158.93	162.50	164.93	167.26	168.93	173.04	174.97
Config	$4f^1 5d^1 6s^2$	$4f^3 6s^2$	$4f^4 6s^2$	$4f^5 6s^2$	$4f^6 6s^2$	$4f^7 6s^2$	$4f^7 5d^1 6s^2$	$4f^8 5d^1 6s^2$	$4f^{10} 6s^2$	$4f^{11} 6s^2$	$4f^{12} 6s^2$	$4f^{13} 6s^2$	$4f^{14} 6s^2$	$4f^{14} 5d^1 6s^2$

Actinides

Property	Th^{90}	Pa^{91}	U^{92}	Np^{93}	Pu^{94}	Am^{95}	Cm^{96}	Bk^{97}	Cf^{98}	Es^{99}	Fm^{100}	Md^{101}	No^{102}	Lr^{103}
Valences	+4	+3 to +5	+3 to +6	+3 to +6	+3 to +6	+3 to +6	+3	+3, +4	+3	+3	+3	+2, +3	+2, +3	+3
Atomic Weight	232.04	231.04	238.03	(237)	(244)	(243)	(247)	(247)	(251)	(254)	(257)	(258)	(259)	(262)
Config	$6d^2 7s^2$	$5f^2 6d^1 7s^2$	$5f^3 6d^1 7s^2$	$5f^4 6d^1 7s^2$	$5f^6 7s^2$	$5f^7 7s^2$	$5f^7 6d^1 7s^2$	$5f^8 6d^1 7s^2$	$5f^{10} 7s^2$	$5f^{11} 7s^2$	$5f^{12} 7s^2$	$5f^{13} 7s^2$	$5f^{14} 7s^2$	$5f^{14} 7s^2 7p^1$

Symbol → Cr^{24} ← Atomic number
+2, +3, +6 ← Valences
52.00 ← Atomic Weight
$3d^5 4s^2$ ← Outer electronic configuration

1
Fundamentals

1.1 THE ORIGINS OF SPECTROSCOPY

The science known as spectroscopy is a branch of physics that deals with the study of the radiation absorbed, reflected, emitted, or scattered by a substance. Although, strictly speaking, the term 'radiation' only deals with photons (electromagnetic radiation), spectroscopy also involves the interactions of other types of particles, such as neutrons, electrons, and protons, which are used to investigate matter.

It is easy to imagine a variety of spectroscopies, as many as the number of possible classifications according to the radiation used and/or the state of the matter (solid, liquid, or gas) interacting with this radiation. The tremendous development of new experimental techniques, as well as the sophistication of those that already exist, is giving rise to the continuous appearance of new spectroscopic techniques. Nevertheless, the different spectroscopies and spectroscopic techniques are rooted in a basic phenomenon: 'the absorption, reflection, emission, or scattering of radiation by matter in a selective range of frequencies and under certain conditions.'

From a historical viewpoint, spectroscopy arose in the 17th century after a famous experiment carried out by Isaac Newton and published in 1672. In this experiment, Newton observed that sunlight contained all the colors of the rainbow, with wavelengths that ranged over the entire visible spectrum (from about 390 nm to 780 nm). He actually labeled this rainbow a 'spectrum.' At the beginning of the 19th century, the spectral range provided by the Newton spectrum was extended with the discovery of new types of electromagnetic radiation that are not visible; infrared (IR) radiation by Herschel (1800), at the long-wavelength end, and ultraviolet (UV) radiation by Ritter (1801) at the short-wavelength end. Both spectral ranges are now of great importance in different areas, such as environmental science (UV and IR) and communications (IR).

An Introduction to the Optical Spectroscopy of Inorganic Solids   J. García Solé, L. E. Bausá, and D. Jaque
© 2005 John Wiley & Sons, Ltd   ISBNs: 0-470-86885-6 (HB); 0-470-86886-4 (PB)

The strong development of optical spectrophotometers during the first half of the 19th century allowed numerous spectra to be registered, such as those of flame colors and the rich line spectra that originate from electrical discharges in atomic gases. With the later development of diffraction gratings, the complicated spectra of molecular gases were analyzed in detail, so that several spectral sharp-line series and new fine structure details were observed. This provided a high-quality step in optical spectroscopy.

For a long time, a large number of spectra were registered but their satisfactory explanation was still lacking. In 1913, the Danish physicist Niels Bohr elaborated a simple theory that allowed an explanation of the hydrogen atomic spectrum previously registered by J. J. Balmer (1885). This constituted a large impulse for the later appearance of quantum mechanics; a fundamental step in interpreting a variety of spectra of atoms and molecules that still lacked a satisfactory explanation.

The interpretation of optical spectra of solids is even more complicated than for atomic and molecular systems, as it requires a previous understanding of their atomic and electronic structure. Unlike liquids and gases, the basic units of solids (atoms or ions) are periodically arranged in long (crystals) or short (glasses) order. This aspect confers particular characteristics to the spectroscopic techniques used to analyze solids, and gives rise to solid state spectroscopy. This new branch of the spectroscopy has led to the appearance of new spectroscopic techniques, which are increasing day by day.

In any case, it is worthwhile to emphasize the dominant role of optical spectroscopy in the investigation of solids. Indeed, the optical spectroscopy of solids appears as a nice 'window' onto the more general field of solid state spectroscopy.

In this chapter, we introduce the fundamentals of the main spectroscopies based on the interaction of optical beams with solid materials.

1.2 THE ELECTROMAGNETIC SPECTRUM AND OPTICAL SPECTROSCOPY

Every day, different types of electromagnetic radiation are invading us; from the low-frequency radiation generated by an AC circuit (≈ 50 Hz) to the highest photon energy radiation of gamma rays (with frequencies up to 10^{22} Hz). These types of radiation are classified according to the *electromagnetic spectrum* (see Figure 1.1), which spans the above-mentioned wide range of frequencies.

The electromagnetic spectrum is traditionally divided into seven well-known spectral regions; radio waves, microwaves, infrared, visible and ultraviolet light, X- (or Roentgen) rays, and γ-rays. All of these radiations have in common the fact that they propagate through the space as transverse electromagnetic waves and at the same speed, $c \cong 3 \times 10^8$ m s^{-1}, in a vacuum.[1] The various spectral regions of the

---

[1] The actual value for the speed of light is given by $c = 2.99792458 \times 10^8$ m s^{-1}.

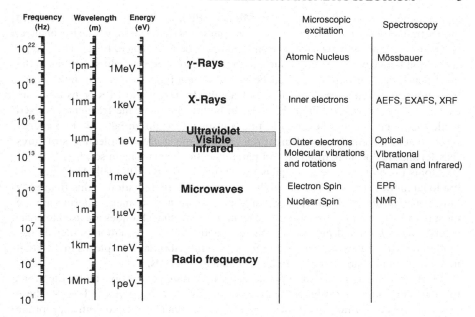

Figure 1.1 The electromagnetic spectrum, showing the different microscopic excitation sources and the spectroscopies related to the different spectral regions. XRF, X-Ray Fluorescence; AEFS, Absorption Edge Fine Structure; EXAFS, Extended X-ray Absorption Fine Structure; NMR, Nuclear Magnetic Resonance; EPR, Electron Paramagnetic Resonance. The shaded region indicates the optical range.

electromagnetic spectrum differ in wavelength and frequency, which leads to substantial differences in their generation, detection, and interaction with matter. The limits between the different regions are fixed by convention rather than by sharp discontinuities of the physical phenomena involved. Each type of monochromatic electromagnetic radiation is usually labeled by its frequency, v, wavelength, λ, photon energy, E, or wavenumber, $\bar{v}$. These magnitudes are interrelated by the well-known quantization equation:

$$E = hv = \frac{hc}{\lambda} = hc\bar{v} \tag{1.1}$$

where $h = 6.62 \times 10^{-34}$ J s is Planck's constant.

The different spectroscopic techniques operate over limited frequency ranges within the electromagnetic spectrum, depending on the processes that are involved and on the magnitudes of the energy changes associated with these processes. Figure 1.1 includes the microscopic entity affected by the excitation, as well as some relevant spectroscopic techniques used in each spectral region.

In magnetic resonance techniques (NMR and EPR), *microwaves* are used in order to induce transitions between different nuclear spin states (Nuclear Magnetic Resonance, NMR) or electron spin states (Electron Paramagnetic Resonance, EPR). The energy separation between the different nuclear spin states or electron spin states lies in the microwave spectral region and can be varied by applying a magnetic field. NMR transitions are excited by frequencies of about 10^8 Hz, while EPR transitions are excited by frequencies of about 10^{10} Hz. In both techniques, the frequency is fixed while the magnetic field is varied to find a resonant condition between two energy levels. These techniques are of great relevance in the study of molecular structures (NMR) and the local environments of paramagnetic dopant ions in solids (EPR).

Atoms in solids vibrate at frequencies of approximately 10^{12}–10^{13} Hz. Thus, vibrational modes can be excited to higher energy states by radiation in this frequency range; that is, *infrared* radiation. Infrared absorption and Raman scattering are the most relevant vibrational spectroscopic techniques. Both techniques are used to characterize the vibrational modes of molecules and solids. Thus, among other things, vibrational techniques are very useful in identifying vibrating complexes in different materials and in characterizing structural changes in solids.

Electronic energy levels are separated by a wide range of energy values. Electrons located in the outer energy levels involve transitions in a range of about 1–6 eV. These electrons are commonly called *valence electrons* and can be excited with appropriate ultraviolet (UV), visible (VIS), or even near infrared (IR) radiation in a wavelength range from about 200 nm to about 3000 nm. This wavelength range is called the *optical range*, and it gives rise to *optical spectroscopy*. Thus, this book is mostly concerned with excitations of valence electrons. These electrons are responsible for a great number of physical and chemical properties; for example, the formation of molecules and solids.

The short (UV) wavelength limit of the optical range is imposed by instrumental considerations (spectrophotometers do not usually work at wavelengths shorter than about 200 nm) and by the validity of the macroscopic Maxwell equations. These equations assume a continuous medium; in other words, that there is a large number of ions within a volume of λ^3. The long (IR) wavelength limit of the optical range is basically imposed by experimental considerations (spectrophotometers work up to about 3000 nm).

Inner electrons are usually excited by *X-rays*. Atoms give characteristic X-ray absorption and emission spectra, due to a variety of ionization and possible inter-shell transitions. Two relevant refined X-ray absorption techniques, that use synchrotron radiation, are the so-called Absorption Edge Fine Structure (AEFS) and Extended X-ray Absorption Fine Structure (EXAFS). These techniques are very useful in the investigation of local structures in solids. On the other hand, X-Ray Fluorescence (XRF) is an important analytical technique.

γ-*Rays* are used in Mössbauer spectroscopy. In some ways, this type of spectroscopy is similar to NMR, as it is concerned with transitions inside atomic nuclei. It provides information on the oxidation state, coordination number, and bond character of specific radioactive ions in solids.

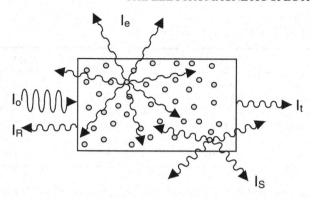

Figure 1.2   The possible emerging beams when a solid sample is illuminated with a beam of intensity I_0. The circles represent atoms or defects in the solid that are interacting with the incoming light.

We shall now focus our attention on the *optical spectroscopy of solids,* which is the subject of this book.

If a solid sample is illuminated by a light[2] beam of intensity I_0, we perceive that, in general, the intensity of this beam is attenuated after crossing the sample; that is, the intensity I_t of the transmitted beam is lower than I_0. The processes that contribute to this attenuation are as follows:

- *Absorption*, if the beam frequency is resonant with a ground to excited state transition of the atoms in the solid. A fraction of this intensity is generally emitted (usually at lower frequency than that of the incident beam), giving rise to an *emission* of intensity I_e. The other fraction of the absorbed intensity is lost by nonradiative processes (heat).

- *Reflection* with an intensity I_R from the external and internal surfaces.

- *Scattering*, with a light intensity I_S spread in several directions, due to elastic (at the same frequency as the incident beam) or inelastic (at lower and higher frequencies than that of the incident beam – Raman scattering) processes.

Figure 1.2 shows the possible emerging beams after an incoming beam of intensity I_0 reaches a solid block. These emerging beams occur as a result of the interaction of the incoming light with atoms and/or defects in the solid: part of the incident intensity is reflected in a backward direction as a beam of intensity I_R. Emitted beams of intensity I_e and/or scattered beams of intensity I_S spread in all directions. The transmitted beam of intensity I_t is also represented.

---

[2] The term 'light' is understood here in a broader sense than visible radiation, as it also includes the UV and IR radiation belonging to the optical range.

Table 1.1 The spectral ranges associated with the different colors perceived by an average person.

Color	λ (nm)	ν (Hz) (× 10^{14})	E (eV)
Violet	390–455	7.69–6.59	3.18–2.73
Blue	455–492	6.59–6.10	2.73–2.52
Green	492–577	6.10–5.20	2.52–2.15
Yellow	577–597	5.20–5.03	2.15–2.08
Orange	597–622	5.03–4.82	2.08–1.99
Red	622–780	4.82–3.84	1.99–1.59

Optical spectroscopy (absorption, luminescence, reflection, and Raman scattering) analyzes the frequency and intensity of these emerging beams as a function of the frequency and intensity of the incident beam. By means of optical spectroscopy, we can understand the color of an object, as it depends on the emission, reflection, and transmission processes of light by the object according to the sensitivity of the human eye to the different colors. The spectral ranges (wavelength, frequency, and photon energy ranges) corresponding to each color for an average person are given in Table 1.1.

EXAMPLE 1.1   *Frequencies and colors of maximum sensitivity for the human eye.*

Vision results from signals transmitted to the brain by about 125 million sensors located in the retina. These photoreceptors are of two types, called *cones* and *rods*. Cones work under intense light – that is, during daylight hours – and this mode of vision is called *photopic vision*. Rods work under dim lighting conditions, and this is called *scotopic vision*.

The relative sensitivities of cones and rods as a function of wavelength are shown in Figure 1.3. There are three types of cones, which contain visual pigments with different spectral sensitivities. The set of cones with a maximum sensibility at 437 nm are especially sensitive (see Table 1.1) to the violet. According to Table 1.1, the cones with a maximum sensibility at 533 nm and 564 nm are both particularly sensitive to green colors, but the latter ones also respond in the red spectral region. Each color results from the combination of three fundamental hues, 'blue,' 'green,' and 'red'; thus it is clear that the cones are responsible for color perception. The cones with a maximum sensitivity at 437 nm are used to detect mainly blue colors, whilst those with maximums at 533 nm and 546 nm are useful in the detection of green and red colors, respectively.

Scotopic vision does not permit the resolution of colors. Rods are much more sensitive to dim light (see Figure 1.3) than cones, but they do not have different types of pigments that are sensitive to different wavelengths; so color cannot be distinguished in dim lighting conditions.

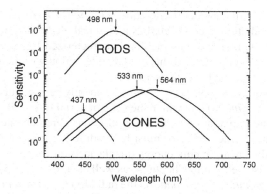

Figure 1.3 The sensitivity of the eye for photopic (cones) and scotopic (rods) vision. The arrows indicate the wavelengths of maximum sensitivity.

We can now estimate the frequency that corresponds to the radiation detected under photopic or scotopic vision. From expression (1.1), we obtain that $\nu = c/\lambda$, so that:

- Rods have a maximum sensitivity to light of

$$\nu = \frac{3 \times 10^{17}\,\text{nm}\,\text{Hz}}{498\,\text{nm}} = 6 \times 10^{14}\,\text{Hz}$$

- The three types of cones have maximum sensitivities at the frequencies 6.8×10^{14} Hz, 5.6×10^{14} Hz, and 5.3×10^{14} Hz.

Optical spectroscopy also provides an excellent tool with which to obtain information on the electronic structure of absorbing/emitting centers (atoms, ions, defects, etc.), their lattice locations, and their environments. In other words, optical spectroscopy allows us to 'look inside' solids by analyzing the emerging light.

Experimental spectra are usually presented as plots of the intensity of (absorbed, emitted, reflected, or scattered) radiation versus the photon energy (in eV), the wavelength (in nm) or the wavenumber (in cm^{-1}). Using Equation (1.1), useful interconversion equations between these different units can be obtained:

$$E\,(\text{eV}) = \frac{1240}{\lambda\,(\text{nm})} = 1.24 \times 10^{-4}\bar{\nu}\,(\text{cm}^{-1})$$

$$\bar{\nu}\,(\text{cm}^{-1}) = \frac{10^7}{\lambda\,(\text{nm})} = 8064.5E\,(\text{eV}) \qquad (1.2)$$

As optical spectroscopy studies the interaction between radiation and matter, three different approximations can be used to account for this interaction.

In Chapter 4 we will consider the so-called *classical approximation*, in which the electromagnetic radiation is considered as a classical electromagnetic wave and the solid is described as a continuous medium, characterized by its relative dielectric constant ε or its magnetic permeability μ. The interaction will then be described by the classical oscillator (the Lorentz oscillator).

However, a variety of spectroscopic features can only be explained by the so-called *semi-classical approximation,* in which the solid material is described by its quantum response, while the propagating radiation is still considered classically. In this case, the classical model of the oscillator must be modified to take into account the fact that solids can only absorb or emit energy quanta according to discrete energy levels. This model will also be used in Chapter 4, as well as in Chapters 5, 6, and 7.

Finally, in the *quantum approximation* the radiation is no longer treated classically (i.e., using Maxwell's equation), and so both radiation and matter are described by quantum methods. For most of the features in the spectra of solids, this approach is not necessary and it will not be invoked. However, this approximation also leads to important aspects, such as zero-point fluctuations, which are relevant in the theory of lasers and Optical Parametic Oscillators (Chapter 3).

1.3 ABSORPTION

1.3.1 The Absorption Coefficient

In the previous section, we have mentioned that a light beam becomes attenuated after passing through a material. Experiments show that the beam intensity attenuation dI after traversing a differential thickness dx can be written as

$$dI = -\alpha I\,dx \tag{1.3}$$

where I is the light intensity at a distance x into the medium and α accounts for the amount of reduction due to the constitution of the material. In the case of negligible scattering, α is called the *absorption coefficient* of the material. Upon integration of Equation (1.3) we obtain

$$I = I_0 e^{-\alpha x} \tag{1.4}$$

which gives an exponential attenuation law relating the incoming light intensity I_0 (the incident intensity minus the reflection losses at the surface) to the thickness x. This law is known as the *Lambert–Beer law*.

From a microscopic point of view of the absorption process, we can assume a simple two energy level quantum system for which N and N' are the ground and excited state population densities (the atoms per unit volume in each state). The absorption coefficient of this system can be written as

$$\alpha(\nu) = \sigma(\nu)(N - N') \tag{1.5}$$

where $\sigma(\nu)$ is the so-called *transition cross section*. For low-intensity incident beams, which is the usual situation in light absorption experiments, $N \gg N'$ and then Equation (1.5) can be written as

$$\alpha(\nu) = \sigma(\nu)N \tag{1.6}$$

where the transition cross section $\sigma(\nu)$ (normally given in cm^2) represents the ability of our system to absorb the incoming radiation of frequency ν. Indeed, the transition cross section is related to the matrix element $| < \Psi_f|H|\Psi_i > |$ of our two-level system, where Ψ_i and Ψ_f denote the eigenfunctions of the ground and excited states, respectively, and H is the interaction Hamiltonian between the incoming light and the system. Equation (1.6) also shows that the absorption coefficient is proportional to the density of absorbing atoms (or centers), N (normally expressed in cm^{-3}).

For our two-level system, we should expect an absorption spectrum like the one displayed in Figure 1.4(a); that is, a delta function at a frequency $\nu_0 = (E_f - E_i)/h$, E_f and E_i being the excited and ground state energies, respectively. However, due

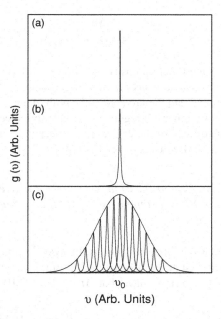

Figure 1.4 (a) An ideal absorption spectrum for a two-level system. (b) The Lorentzian line shape of an optical band, related to homogeneous broadening. (c) The Gaussian line shape of an optical band, related to inhomogeneous broadening.

to various line-broadening mechanisms, the observed spectrum never consists of a single line, but of a band.

In fact, the transition cross section can be written in terms of a *line-shape function* $g(v)$ (with units of Hz^{-1}) in the following way:

$$\sigma(v) = S \times g(v) \tag{1.7}$$

where $S = \int_0^\infty \sigma(v)\,dv$ is the so-called *transition strength* and represents the full strength of the transition to absorb (or emit) radiation.

The line-shape function gives the profile of the optical absorption (and emission) band and contains important information about the photon–system interaction. Let us briefly discuss the different mechanisms that contribute to this function, or the different *line-broadening mechanisms*.

The ultimate (minimum) linewidth of an optical band is due to the *natural or lifetime broadening*. This broadening arises from the Heisenberg's uncertainty principle, $\Delta v \Delta t \leq 1/2\pi$, Δv being the full frequency width at half maximum of the transition and Δt the time available to measure the frequency of the transition (basically, the lifetime of the excited state, as defined in the next section). This broadening mechanism leads to a Lorentzian profile given by (Svelto, 1986):

$$g(v) = \frac{\Delta v/2\pi}{(v - v_0)^2 + (\Delta v/2)^2} \tag{1.8}$$

This Lorentzian line-shape function has been sketched in Figure 1.4(b). The natural broadening is a type of *homogeneous broadening*, in which all the absorbing atoms are assumed to be identical and then to contribute with identical line-shape functions to the spectrum. There are other homogeneous broadening mechanisms, such as that due to the dynamic distortions of the crystalline environment associated with lattice vibrations, which are partially discussed in Chapter 5.

EXAMPLE 1.2   *An allowed emission transition for a given optical ion in a solid has a lifetime of 10 ns. Estimate its natural broadening. Then estimate the peak value of $g(v)$.*

Considering that this lifetime value is the average time for which the ion remains in the excited state (and hence is the time available to measure the frequency of the transition), we have that $\Delta t \approx 10\,\text{ns} = 10^{-8}\,\text{s}$, and so

$$(\Delta v)_{\text{nat}} \approx \frac{1}{2\pi\,\Delta t} = 1.6 \times 10^7\,\text{Hz} = 16\,\text{MHz}$$

Due to natural broadening, this ion would present a linewidth function given by Equation (1.8), with a full width at half maximum of $\Delta v = 16\,\text{MHz}$.

This broadening is much less than is actually observed for optical ions in solids. In fact, natural broadening can only be resolved in some solids and at very low temperatures, so that the atoms are rigidly fixed and do not interact with other ions (negligible dynamic distortions).

Now using Equation (1.8) and $\Delta v = 16$ MHz, we can determine the peak value for the line-shape function:

$$g(v_0) = \frac{2}{\pi \Delta v} = 4 \times 10^{-8} \, \text{Hz}^{-1}$$

In various cases, the different absorbing centers have different resonant frequencies, so that the line shape results from the convolution of the line shapes of the different centers, weighted by their corresponding concentrations, as shown in Figure 1.4(c). This type of broadening is called *inhomogeneous broadening* and, in general, it leads to a Gaussian line shape, given by the expression (Svelto, 1986)

$$g(v) = \frac{2}{\Delta v} \left(\frac{\ln 2}{\pi} \right)^{1/2} e^{-\left(\frac{v - v_0}{\Delta v / 2} \right)^2 \times \ln 2} \qquad (1.9)$$

Inhomogeneous broadening in solids typically occurs as a result of nonequivalent static distortions in the crystalline environment of an optically active center. As can happen with the paving stones in a floor, the crystal reticules are not perfectly equal; there is a distribution of crystalline environments for the absorbing atom, and consequently a distribution of resonance frequencies.

The line-shape function of a given transition informs us on the particular character of the interaction of the absorbing atom with its environment in the solid. In the most general case, this line shape is due to the combined effect of more than one independent broadening mechanism. In this case, the overall line shape is given by the convolution of the line-shape functions associated with the different broadening mechanisms.

Usually, optical bands of centers (atoms, ions, etc.) in solids are broader than those found in gases or liquids. This is because centers are in general more diluted (isolated) in gases or liquids than in solids, and so they are subjected to less important interactions with their neighborhood.

1.3.2 The Measurement of Absorption Spectra: the Spectrophotometer

Absorption spectra are usually registered by instruments known as *spectrophotometers*. Figure 1.5(a) shows a schematic diagram with the main elements of the simplest spectrophotometer (*a single-beam spectrophotometer*). Basically, it consists of the following elements: (i) a light source (usually a deuterium lamp for the UV spectral

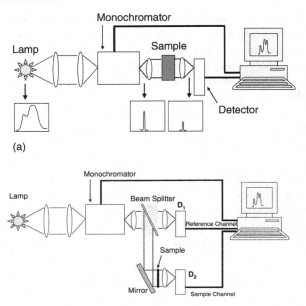

Figure 1.5 Schematic diagrams of (a) a single-beam spectrophotometer and (b) a double-beam spectrophotometer.

range and a tungsten lamp for the VIS and IR spectral ranges) that is focused on the entrance to (ii) a monochromator, which is used to select a single frequency (wavelength) from all of those provided by the lamp source and to scan over a desired frequency range; (iii) a sample holder, followed by (iv) a light detector (usually a photomultiplier for the UV–VIS range and a SPb cell for the IR range) to measure the intensity of each monochromatic beam after traversing the sample; and finally (v) a computer, to display and record the absorption spectrum.

The spectral shape of the light after passing through each element has also been sketched in Figure 1.5(a). Details of how these components work are given in the next chapter.

Optical spectrophotometers work in different modes to measure *optical density*, *absorbance*, or *transmittance*.

The *optical density* is defined as $OD = \log(I_0/I)$, so that according to Equation (1.4) the absorption coefficient is determined by

$$\alpha = \frac{(OD)}{x \log e} = \frac{2.303(OD)}{x} \tag{1.10}$$

That is, by measuring the optical density and the sample thickness, the absorption coefficient can be determined. According to Equation (1.6) we can now determine the absorption cross section σ if the density of centers is known. This means that, by a simple absorption spectrum, information about the matrix element $|< \Psi_f|H|\Psi_i >|$ of our quantum system can be obtained. On the other hand, if σ is known, the concentration of absorbing centers, N, can be estimated.

The optical density can be easily related to other well-known optical magnitudes that are also directly measurable by spectrophotometers, such as the *transmittance*, $T = I/I_0$, and the *absorbance*,[3] $A = 1 - I/I_0$:

$$T = 10^{-OD}$$
$$A = 1 - 10^{-OD} \qquad (1.11)$$

Nevertheless, it is important to emphasize here the advantage of measuring optical density spectra over transmittance or absorbance spectra. Optical density spectra are more sensitive, as they provide a higher contrast than absorbance or transmittance spectra (as can be seen after solving Exercise 1.2). In fact, for low optical densities, expression (1.11) gives $A \approx 1 - (1 - OD) = OD$, so that the absorbance spectrum (A versus λ, or $1 - T$ versus λ) displays the same shape as the optical density. However, for high optical densities, typically higher than 0.2 (see Exercise 1.2), the absorbance spectrum gives a quite different shape to that of the actual absorption (α versus λ or OD versus λ) spectrum.

Another optical absorption magnitude, which is of relevance when working with optical fibers, is the *attenuation*. Exercise 1.4 deals with this magnitude.

A single-beam spectrophotometer, like the one shown in Figure 1.5(a), presents a variety of problems, because the spectra are affected by spectral and temporal variations in the illumination intensity. The spectral variations are due to the combined effects of the lamp spectrum and the monochromator response, while the temporal variations occur because of lamp stability.

To reduce these effects, *double-beam spectrophotometers* are used. Figure 1.5(b) shows a schematic diagram of the main components of these spectrophotometers. The illuminating beam is split into two beams of equal intensity, which are directed toward two different channels; a reference channel and a sample channel. The outgoing intensities correspond respectively to I_0 and I, which are detected by two similar detectors, D_1 and D_2 in Figure 1.5(b). As a consequence, the spectral and temporal intensity variations of the illuminating beam affect both the reference and sample beams in the same way, and these effects are minimized in the resulting absorption spectrum. The system can still be improved if a unique detector is used instead of the two detectors, D_1 and D_2. In such a way, the errors introduced by the usual nonexactly equal spectral responses of these detectors are also eliminated. This can be achieved by introducing fast rotating mirrors so that the intensities I_0 and I can be always sent to the same detector.

Typical sensitivities of $(OD)_{min} \approx 5 \times 10^{-3}$ can be reached with these spectrophotometers.

---

[3] Recall that I_0 is the intensity of the light entering the sample; that is, without reflection.

EXAMPLE 1.3 *If the cross section for a given transition of Nd^{3+} ions in a particular crystal is 10^{-19} cm^2 and a sample of thickness 0.5 mm is used, determine the minimum concentration of absorbing ions that can be detected with a typical double-beam spectrophotometer.*

For a typical double-beam spectrophotometer, the sensitivity in terms of the optical density is $(OD)_{min} \approx 5 \times 10^{-3}$. Therefore, using Equations (1.6) and (1.10), the minimum concentration of absorbing centers that can be detected is

$$N_{min} = \frac{2.303(OD)_{min}}{x\sigma} = \frac{2.303 \times 5 \times 10^{-3}}{0.05 \times 10^{-19}} \approx 10^{18}\ cm^{-3}$$

Crystals have typical constituent ion concentrations of about 10^{22} cm^{-3}, so that the previous minimum concentration of Nd^{3+} ions corresponds to about 0.01 % or 100 parts per million (ppm).

The previous sensitivity can be still improved by calculating the second derivative spectrum; that is, $d^2(OD)/d\lambda^2$ versus λ. Figure 1.6 is a nice example of this improvement. It shows the room temperature absorption spectrum of a natural colorless

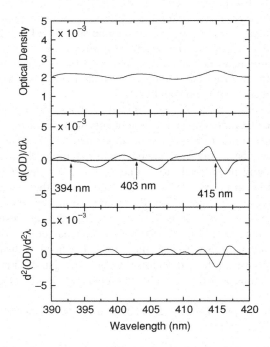

Figure 1.6 The absorption spectrum and first and second derivative spectra of a natural colorless diamond. The marked wavelengths indicate characteristic bands of N_3 centers (reproduced with permission from Lifante *et al.*, 1990).

diamond together with the first and second derivative spectra. The features around 394 nm, 403 nm, and 415 nm in the derivative spectra (a single inflexion point in the first derivative spectrum must correspond to a minimum in the second derivative spectrum) correspond to vibrational bands associated with the so-called N_3 centers, which are characteristic of the natural gemstone. These features are practically unobservable in the absorption spectrum. Indeed, this improvement due to the data treatment of optical spectra is very useful in gemmology.

Finally, mention should be made of the new spectrophotometers based on an array of detectors (see Chapter 2). The operation of these spectrophotometers is much simpler, as they are not monochromators, each detector in the array being used to record a wavelength interval. However, these monochromators suffer from the drawback of poor spectral resolution.

1.3.3 Reflectivity

Reflectivity spectra provide similar and complementary information to the absorption measurements. For instance, absorption coefficients corresponding to the fundamental absorption (see Chapter 4) are as high as 10^5–10^6 cm^{-1}, so that they can only be measured by using very thin samples (thin films). In these cases, the reflectivity spectra $R(v)$ can be very advantageous, as they manifest the singularities caused by the absorption process but with the possibility of using bulk samples. In fact, the reflectivity, $R(v)$, and the absorption spectra, $\alpha(v)$, can be interrelated by using the so-called Kramers–Krönig relations (Fox, 2001).

The reflectivity at each frequency is defined by

$$R = \frac{I_R}{I_0} \tag{1.12}$$

where I_R is the reflected intensity.

Reflectivity spectra can be registered in two different modes: (i) *direct reflectivity* or (ii) *diffuse reflectivity*. Direct reflectivity measurements are made with well-polished samples at normal incidence. Diffuse reflectivity is generally used for unpolished or powdered samples. Figure 1.7 shows the experimental arrangements for measuring both types of spectra.

(i) For direct reflectivity measurements (Figure 1.7(a)), monochromatic light (produced by a lamp and a monochromator) is passed through a semitransparent lamina (the beam splitter in Figure 1.7(a)). This lamina deviates the light reflected in the sample toward a detector.

(ii) For diffuse reflectivity measurements, an integrating sphere (a sphere with a fully reflective inner surface) is used (Figure 1.7(b)). Such a sphere has a pinhole through which the light enters and is transmitted toward the sample. The diffuse reflected light reaches the detector after suffering multiple reflections in the inner surface of the sphere. The integrating spheres can be incorporated as additional instrumentation into conventional spectrophotometers.

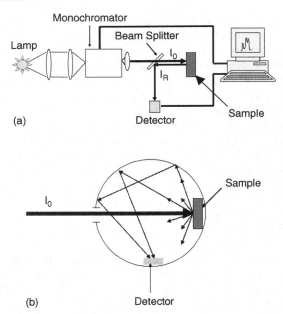

Figure 1.7 (a) An experimental arrangement used to measure direct reflectivity spectra. (b) A schematic drawing of an integrating sphere for measuring diffuse reflectivity spectra.

1.4 LUMINESCENCE

Luminescence is, in some ways, the inverse process to absorption. We have seen in the previous section how a simple two-level atomic system shifts to the excited state after photons of appropriate frequency are absorbed. This atomic system can return to the ground state by spontaneous emission of photons. This de-excitation process is called *luminescence*. However, the absorption of light is only one of the multiple mechanisms by which a system can be excited. In a general sense, luminescence is the emission of light from a system that is excited by some form of energy. Table 1.2 lists the most important types of luminescence according to the excitation mechanism.

Photoluminescence occurs after excitation with light (i.e., radiation within the optical range). For instance, several organic laundry detergents make use of photoluminescence dyes. In fact, the slogan 'it washes more white,' used with some detergents, could be substituted by 'it is more luminescent.' Luminescence can also be produced under excitation with an electron beam, and in this case it is called *cathodoluminescence*. This technique is conventionally used to investigate some characteristics of specimens, such as trace impurities and lattice defects, as well as to investigate crystal distortion. Excitation by high-energy electromagnetic radiation (sometimes called ionizing radiation) such as X-rays, α-rays (helium nuclei), β-rays (electrons), or γ-rays leads to a type of photoluminescence called *radioluminescence*. Scintillation

Table 1.2 The various types of luminescence

Name	Excitation mechanism
Photoluminescence	Light
Cathodoluminescence	Electrons
Radioluminescence	X-rays, α-, β-, or γ-rays
Thermoluminescence	Heating
Electroluminescence	Electric field or current
Triboluminescence	Mechanical energy
Sonoluminescence	Sound waves in liquids
Chemiluminescence and bioluminescence	Chemical reactions

counters are based on this luminescence mechanism; a crystal is excited by energetic rays (radioactive radiation) and produces a luminescence that is detected by a photomultiplier. *Thermoluminescence* occurs when a substance emits light as a result of the release of energy stored in traps by thermal heating. This mechanism is different to thermally produced blackbody radiation. Thermoluminescence is used, for instance, for dating minerals and ancient ceramics. *Electroluminescence* occurs as a result of the passage of an electric current through a material, as in nightlight panels. *Triboluminescence* is the production of light by a mechanical disturbance, such as the light that emerges when some adhesive tapes are unrolled. Acoustic waves (sound) passing through a liquid can produce *sonoluminescence*. *Chemiluminescence* appears as a result of a chemical reaction. It is used, for instance, in the detection and concentration measurements of some atmosphere contaminants, such as NO_2 and NO. As a particular class of chemiluminescence, we can mention *bioluminescence*, which occurs as a result of chemical reactions inside an organism. Bacteria, jellies, algae, and other organisms, such as fish and squids, are able to produce light by chemicals that they have stored in their bodies. Bioluminescence is the predominant source of light in the deep ocean.

1.4.1 The Measurement of Photoluminescence: the Spectrofluorimeter

We will now focus our attention on the photoluminescence process. A typical experimental arrangement to measure photoluminescence spectra is sketched in Figure 1.8. Photoluminescence spectra are also often measured using compact commercial equipment called *spectrofluorimeters*. Their main elements are also shown in Figure 1.8.

The sample is excited with a lamp, which is followed by a monochromator (the excitation monochromator) or a laser beam. The emitted light is collected by a focusing lens and analyzed by means of a second monochromator (the emission

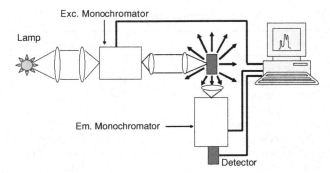

Figure 1.8   A schematic diagram showing the main elements for measuring photoluminescence spectra. The excitation can also be produced using a laser instead of both a lamp and an excitation monochromator.

monochromator), followed by a suitable detector connected to a computer. Two kinds of spectra, (i) *emission spectra* and (ii) *excitation spectra*, can be registered:

(i) In *emission spectra*, the excitation wavelength is fixed and the emitted light intensity is measured at different wavelengths by scanning the emission monochromator.

(ii) In *excitation spectra*, the emission monochromator is fixed at any emission wavelength while the excitation wavelength is scanned in a certain spectral range.

The differences between excitation and emission spectra can be better understood after a careful reading of the next example.

EXAMPLE 1.4   *Consider a phosphor with a three energy level scheme and the absorption spectrum shown in Figure 1.9(a). Assuming similar transition probabilities among these levels, discuss the nature of the excitation and emission spectra and their relationship to the absorption spectrum.*

The absorption spectrum of Figure 1.9(a) shows two bands at photon energies $h\nu_1$ and $h\nu_2$, corresponding to the $0 \rightarrow 1$ and $0 \rightarrow 2$, transitions respectively.

Let us first discuss on the possible emission spectra. Excitation with light of energy $h\nu_1$ promotes electrons from the ground state 0 to the excited state 1, which becomes populated. Thus, as shown in Figure 1.9(b), the emission spectrum when fixing the excitation energy at $h\nu_1$ consists of a single band that peaks at this same photon energy, $h\nu_1$. On the other hand, when the excitation energy is fixed at $h\nu_2$, the state 2 is populated and an emission spectrum like the one shown in Figure 1.9(c) can be produced. This emission spectrum shows three bands that peak at energies of $h(\nu_2 - \nu_1)$, $h\nu_1$, and $h\nu_2$, related to the transitions $2 \rightarrow 1$, $1 \rightarrow 0$, and $2 \rightarrow 0$, respectively.

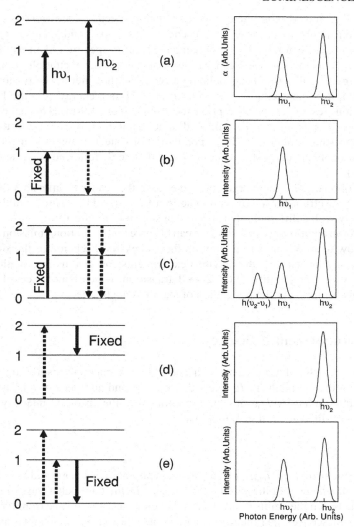

Figure 1.9 The energy-level and transition schemes and possible luminescence spectra of a three-level ideal phosphor: (a) the absorption spectrum; (b, c) emission spectra under excitation with light of photon energies $h\nu_1$ and $h\nu_2$, respectively; (d, e) Excitation spectra monitoring emission energies at $h(\nu_2 - \nu_1)$ and at $h\nu_1$, respectively. Arrows mark the absorption/emission transitions involved in each spectrum. 'Fixed' indicates that the excitation or emission monochromator is fixed at the energy (wavelength) corresponding to this transition.

Let us now discuss the different excitation spectra. If we were to set the emission monochromator at a fixed energy $h(\nu_2 - \nu_1)$ and scan the excitation monochromator, we would obtain an excitation spectrum like that of Figure 1.9(d), as this emission can be only obtained under excitation with light of energy $h\nu_2$; that is, after populating level 2. On the other hand, as shown in Figure 1.9(e), the excitation spectrum for an emission energy $h\nu_1$ (the $1 \rightarrow 0$ transition) resembles the absorption spectrum (Figure 1.9(a)). This is because the $1 \rightarrow 0$ emission can be produced under excitation in any one of the absorption bands; either by a direct population of state 1 or by an indirect population of this excited state (via the excited state 2, followed by a decay to state 1).

Following the same arguments, we leave the reader to infer that the excitation spectrum for an emission energy of $h\nu_2$ would be similar (although probably with a different intensity) to that shown in Figure 1.9(d).

As an additional conclusion, we can observe that excitation/emission spectra allow us to obtain information on the energy-level scheme of the studied phosphor. For instance, the previous experiments can give us information on the transition probability of the $2 \leftrightarrow 1$ transition, which is not accessible by optical absorption as the population of state 1 is negligible.

1.4.2 Luminescent Efficiency

We know that photoluminescence can occur after a material absorbs light. Thus, considering that an intensity I_0 enters the material and an intensity I passes out of it, the emitted intensity I_{em} must be proportional to the absorbed intensity; that is, $I_{em} \propto I_0 - I$. In general, it is written that

$$I_{em} = \eta(I_0 - I) \qquad (1.13)$$

where the intensities, I_0, I_{em}, and I, are given in photons per second and η is called the *luminescent efficiency* or the *quantum efficiency*.[4] Defined in such a way, η represents the ratio between the emitted and absorbed photons and it can vary from 0 to 1.

In a luminescence experiment, only a fraction of the total emitted light is measured. This fraction depends on the focusing system and on the geometric characteristics of the detector. Therefore, in general, the measured emitted intensity, (I_{em}), can be written in terms of the incident intensity I_0 as[5]

$$(I_{em}) = k_g \times \eta \times I_0(1 - 10^{-(OD)}) \qquad (1.14)$$

where k_g is a geometric factor that depends on the experimental setup (the arrangement of the optical components and the detector size) and OD is the optical density of the

---

[4] Formally speaking, these are not units of light intensity (energy per second per unit area) but there is no loss of generality. We will realize the need of using these units below, when defining the Stokes shift.

[5] Note the different notation used for the measured intensity, (I_{em}), and for the total emitted intensity, I_{em}.

sample. For low optical densities, Equation (1.14) becomes

$$(I_{em}) \cong k_g \times \eta \times I_0 \times (OD) \tag{1.15}$$

It is clear from Equation (1.15) that the emitted intensity is linearly dependent on the incident intensity and is proportional to both the quantum efficiency and the optical density (this only for low optical densities). A quantum efficiency of $\eta < 1$ indicates that a fraction of the absorbed energy is lost by *nonradiative* processes. Normally, these processes (which are discussed in Chapters 5 and 6) lead to sample heating. The proportionality to OD, which only holds for low optical densities, indicates that the excitation spectra only reproduce the shape of absorption spectra for samples with low concentrations.

EXAMPLE 1.5   *The sensitivity of luminescence. Consider a photoluminescence experiment in which the excitation source provides a power of 100 μW at a wavelength of 400 nm. The phosphor sample can absorb light at this wavelength and emit light with a quantum efficiency of $\eta = 0.1$. Assuming that $k_g = 10^{-3}$ (i.e., only one-thousandth of the emitted light reaches the detector) and a minimum detectable intensity of 10^3 photons per second, determine the minimum optical density that can be detected by luminescence.*

Each incident photon has an energy of

$$hc/\lambda = \frac{6.62 \times 10^{-34} \text{ J s} \times 3 \times 10^8 \text{ m s}^{-1}}{400 \times 10^{-9} \text{ m}} = 4.96 \times 10^{-19} \text{ J}.$$

Therefore, the incident intensity is

$$I_0 = \frac{10^{-4} \text{ W}}{4.96 \times 10^{-19} \text{ J}} = 2 \times 10^{14} \text{ photons s}^{-1}.$$

The minimum optical density, $(OD)_{min}$, detectable with our experimental setup can be obtained from expression (1.15):

$$(OD)_{min} = \frac{(I_{em})_{min}}{\eta I_0 k_g} = \frac{10^3 \text{ photons s}^{-1}}{0.1 \times 2 \times 10^{14} \text{ photons s}^{-1} \times 10^{-3}} = 5 \times 10^{-8}$$

Comparing this value to the typical sensitivity provided by a spectropho-tometer, $(OD)_{min} = 5 \times 10^{-3}$ (see Section 1.4), we see that the luminescence technique is much more sensitive than the absorption technique (about 10^5 times for this experiment). Although this large sensitivity is an advantage of photoluminescence, care must be taken, as signals from undesired trace lumi-nescent elements (not related to our luminescent center) can overlap with our luminescent signal.

(a)   (b)

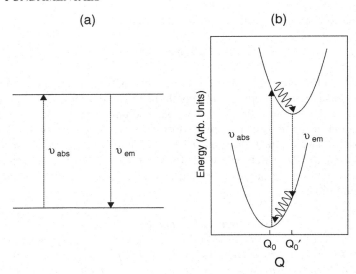

Figure 1.10   (a) The absorption and emission energies for a two-level system (rigid lattice). (b) The absorption and emission energies showing the Stokes shift (vibrating lattice).

1.4.3 Stokes and Anti-Stokes Shifts

Up to now, we have considered that, for our single two-level system, the absorption and emission spectra peak at the same energy, in accordance with the energy levels diagram of Figure 1.10(a). In general, this is not true, and the emission spectrum is shifted to lower energies relative to the absorption spectrum. This shift is called a *Stokes shift* and, although it will be discussed in detail in Section 5.5, we can now give a simple explanation for it. Let us consider that the two-level system of Figure 1.10(a) corresponds to an optical ion embedded in an ionic crystal. These two energy levels are a consequence of the optical ion and its neighbor's ions being at fixed positions (rigid lattice). However, we know that ions in solids are vibrating around equilibrium positions, so that our optical ion sees the neighbors at different distances, oscillating around equilibrium positions. Consequently, each state of Figure 1.10(a) must be now considered as a continuum of states, related to the possible optical ion – neighbor ion distances. Assuming a single coordinate distance Q, and that neighboring ions follow a harmonic motion, the two energy levels of Figure 1.10(a) become parabolas, as in Figure 1.10(b) (according to the potential energy of the harmonic oscillator).

In the spirit of this approach, in Section 5.4 we will justify our claim that the equilibrium positions of the ground and excited states can be different and that the electronic transitions occur as shown in Figure 1.10(b). Four steps can be considered. First, an electron in the ground state is promoted to the excited state without any change in Q_0 (the equilibrium position in the ground state). Afterwards, the electron relaxes within the electronic state to its minimum position Q'_0, its equilibrium position in the excited state. This relaxation is a nonradiative process accompanied by phonon emissions. From this minimum Q'_0, fluorescence is produced from the excited state to the ground state, without any change in the distance coordinate, $Q = Q'_0$. Finally, the electron relaxes within the electronic state to the minimum of the ground state, Q_0. As a result of these four processes, the emission occurs at a frequency ν_{em}, which is lower than ν_{abs}. The energy difference $\Delta = h\nu_{\text{abs}} - h\nu_{\text{em}}$ is a measure of the Stokes shift.

Once the Stokes shift has been introduced, we can better understand the definition of quantum efficiency in terms of absorbed and emitted photons per second rather than the absorbed and emitted intensity (the energy per second per unit area). In fact, it is possible to have a system for which $\eta = 1$ but, because of the Stokes shift, the radiative emitted energy can be lower than the absorbed energy. The fraction of the absorbed energy that is not emitted is delivered as phonons (heat) to the crystal lattice.

It is also possible to obtain luminescence at photon energies higher than the absorbed photon energy. This is called *anti-Stokes or up-conversion luminescence* and it occurs for multilevel systems, as in the example shown in Figure 1.11. For this

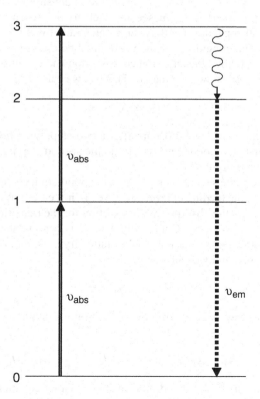

Figure 1.11   An example of a four energy level system producing anti-Stokes luminescence.

system, two photons of frequency ν_{abs} are sequentially absorbed from the ground state 0 and then from the first excited state 1, thus promoting an electron to the excited state 3. Then, the electron decays nonradiatively to state 2, from which the anti-Stokes luminescence $2 \to 1$ is produced. We thus observe how $\nu_{abs} < \nu_{em}$ (the anti-Stokes shift).

As we will see in the next example, anti-Stokes luminescence is, in general, a nonlinear process that is at variance with the normal luminescence process for which, according to Equation (1.15), the emission intensity is proportional to the light excitation intensity I_0.

> EXAMPLE 1.6   *Determine how the emission intensity depends on the excitation intensity for the anti-Stokes luminescence of the system shown in Figure 1.11.*
>
> Let N_0, N_1, N_2, and N_3 be the equilibrium population densities of the states 0, 1, 2, and 3, respectively (reached under continuous wave excitation intensity I_0), and let $N = N_0 + N_1 + N_2 + N_3$ be the total density of optical absorbing centers. The up-converted luminescence intensity I_{20} (corresponding to the transition $2 \to 0$) depends on both N_2 and on the radiative emission probability of level 2, A_2. This magnitude, which is defined below, is proportional to the cross section σ_{20} (called the emission cross section and equal to the absorption cross section σ_{02}, as shown in Chapter 5). Thus we can write
>
> $$I_{20} \propto N_2 \times A_2$$
>
> The population of level 2 occurs after a two-step, two-photon absorption process. The first step populates level 1 and the second step populates level 3, which feeds the emitting level 2.
>
> As a result of the second step process, the population of the emitting state 2 ($N_3 \approx 0$, due to the feeding of level 2 by level 3), increases with the population of level 1, with σ_{13} (the absorption cross section for the transition $1 \to 3$), and with the excitation intensity. Conversely, level 2 is depopulated as a result of the up-converted $2 \to 0$ emission. In the stationary situation, both population rates must be equal and we can write
>
> $$N_2 \times A_2 = N_1 \times \sigma_{13} \times I_0$$
>
> In a similar way, we can write the equilibrium equation related to the first step:
>
> $$N_1 \times A_1 = N_0 \times \sigma_{01} \times I_0 - N_1 \times \sigma_{13} \times I_0$$
>
> where the term on the left-hand side in the previous equation accounts for the emission from level 1 (A_1 being the radiative emission probability from

level 1) and the two right-hand terms (σ_{01} being the absorption cross section of transition $0 \rightarrow 1$) are the pump filling ($N_0 \times \sigma_{01} \times I_0$) and pump depopulating ($N_1 \times \sigma_{13} \times I_0$) rates. For not very high intensities $N_1 \ll N_0$ and, we can neglect the second term on the right-hand side of the previous equation, so that $N_1 \propto N_0 \times I_0$. Therefore, $N_2 \propto N_1 \times I_0 \propto N_0 \times I_0^2$. Preserving the fact that the illumination intensities are not very high, we can make the approximation $N \cong N_0$ and then write

$$I_{20} \propto N \times I_0^2$$

As the total density of absorbing centers remains constant with the excitation intensity, this result shows that the intensity of the anti-Stokes luminescence, I_{20}, varies with the square of the excitation intensity.

For higher-order absorption processes, the intensity of the anti-Stokes luminescence also depends on higher powers of the excitation intensity (see Exercise 1.6).

1.4.4 Time-Resolved Luminescence

In the previous sections we have considered that the excitation intensity is kept constant at each wavelength; that is, we have been dealing with *continuous wave excitation*. This situation corresponds to the stationary case (stationary optical excitation), in which the optical feeding into the excited level equals the decay rate to the ground state and so the emitted intensity remains constant with time. Relevant information can be obtained under *pulsed wave excitation*. This type of excitation promotes a nonstationary density of centers N in the excited state. These excited centers can decay to the ground state by *radiative* (light-emitting) and *nonradiative* processes, giving a decay-time intensity signal. The temporal evolution of the excited state population follows a very general rule:

$$\frac{dN(t)}{dt} = -A_T N(t) \tag{1.16}$$

where A_T is the total decay rate (or total decay probability), which is written as:

$$A_T = A + A_{nr} \tag{1.17}$$

A being the *radiative rate* (labeled in such a way because it coincides with the Einstein coefficient of spontaneous emission) and A_{nr} being the *nonradiative rate*; that is, the rate for nonradiative processes. The solution of the differential equation (1.16) gives the density of excited centers at any time t:

$$N(t) = N_0 e^{-A_T t} \tag{1.18}$$

where N_0 is the density of excited centers at $t = 0$; that is, just after the pulse of light has been absorbed.

The de-excitation process can be experimentally observed by analyzing the temporal decay of the emitted light. In fact, the emitted light intensity at a given time t, $I_{em}(t)$, is proportional to the density of centers de-excited per unit time, $(dN/dt)_{radiative} = AN(t)$, so that it can be written as

$$I_{em}(t) = C \times AN(t) = I_0 e^{-A_T t} \tag{1.19}$$

where C is a proportionality constant and so $I_0 = C \times AN_0$ is the intensity at $t = 0$. Equation (1.19) corresponds to an exponential decay law for the emitted intensity, with a *lifetime* given by $\tau = 1/A_T$. This lifetime represents the time in which the emitted intensity decays to I_0/e and it can be obtained from the slope of the linear plot, log I versus t. As τ is measured from a pulsed luminescence experiment, it is called the *fluorescence or luminescence lifetime*. It is important to stress that this lifetime value gives the total decay rate (radiative plus nonradiative rates). Consequently Equation (1.17) is usually written as

$$\frac{1}{\tau} = \frac{1}{\tau_0} + A_{nr} \tag{1.20}$$

where $\tau_0 = 1/A$, called the *radiative lifetime*, would be the luminescence decay time measured for a purely radiative process ($A_{nr} = 0$). In the general case $\tau < \tau_0$, as the nonradiative rate differs from zero.

The quantum efficiency η can now easily be expressed in terms of the radiative τ_0 and luminescence τ lifetimes:

$$\eta = \frac{A}{A + A_{nr}} = \frac{\tau}{\tau_0} \tag{1.21}$$

The previous formula indicates that the radiative lifetime τ_0 (and hence the radiative rate A) can be determined from luminescence decay-time measurements if the quantum efficiency η is measured by an independent experiment. Methods devoted to the measurement of quantum efficiencies are given in Section 5.7.

EXAMPLE 1.7   *The fluorescence lifetime measured from the metastable state* $^4F_{3/2}$ *of* Nd^{3+} *ions in the laser crystal yttrium aluminum borate* ($YAl_3(BO_3)_4$) *is 56 μs. If the quantum efficiency from this state is 0.26, determine the radiative lifetime and the radiative and nonradiative rates.*

According to Equation (1.21), the radiative lifetime τ_0 is

$$\tau_0 = \frac{56 \ \mu s}{0.26} = 215.4 \ \mu s$$

and therefore the radiative rate is given by

$$A = \frac{1}{215.4 \times 10^{-6} \ s} = 4643 \ s^{-1}$$

The total de-excitation rate is

$$A_T = \frac{1}{\tau} = \frac{1}{56 \times 10^{-6} \text{ s}} = 17857 \text{ s}^{-1}$$

so that the nonradiative rate is

$$A_{nr} = A_T - A = 13214 \text{ s}^{-1}$$

The nonradiative rate A_{nr} is much larger than the radiative rate A; as a result, noticeable pump-induced heating effects occur in this laser crystal (Jaque *et al.*, 2000).

At this point it is important to mention that the experimental setup used for luminescence decay-time measurements is similar to that of Figure 1.8, although the light source must be pulsed (alternatively, a pulsed laser can be used) and the detector must be connected to a time-sensitive system, such as an oscilloscope, a multichannel analyzer, or a boxcar integrator (see Chapter 2).

The emission spectra can also be recorded at different times after the excitation pulse has been absorbed. This experimental procedure is called *time-resolved luminescence* and may prove to be of great utility in the understanding of complicated emitting systems. The basic idea of this technique is to record the emission spectrum at a certain *delay time*, t, in respect to the excitation pulse and within a temporal *gate*, Δt, as schematically shown in Figure 1.12. Thus, for different delay times different spectral shapes are obtained.

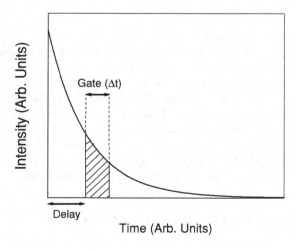

Figure 1.12   The schematic temporal decay of luminescence, showing a gate of width Δt at a delay time t.

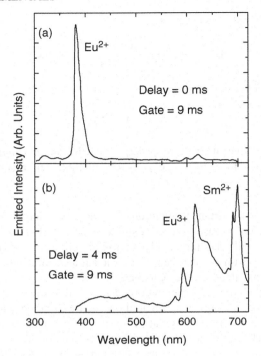

Figure 1.13   The time-resolved emission spectrum of anhydrite ($CaSO_4$) at two different delay times. The scale on the emission intensity axis of (b) has been enlarged by a factor of 1000 in order to clearly observe the remaining luminescence of the Eu^{3+} and Sm^{3+} ions. Excitation wavelength at 266 nm (reproduced with permission from Gaft *et al.*, 2001).

Among other examples, time-resolved luminescence has recently been applied to the detection of different trace elements (i.e., elements in very low concentrations) in minerals. Figure 1.13 shows two time-resolved emission spectra of anhydrite ($CaSO_4$). The emission spectrum just after the excitation pulse (delay 0 ms) shows an emission band peaking at 385 nm, characteristic of Eu^{2+} ions. When the emission spectrum is taken 4 ms after the pulse, the Eu^{2+} luminescence has completely disappeared, as this luminescence has a lifetime of about $10\,\mu s$. This allows us to observe the weak emission signals of the Eu^{3+} and Sm^{3+} ions present in this mineral, which in short time intervals are masked by the Eu^{2+} luminescence. The trivalent ions have larger lifetimes and their luminescence still remains in the ms delay range.

1.5 SCATTERING: THE RAMAN EFFECT

Let us deal with the study of the fraction of light scattered from incident light (I_S in Figure 1.2). A common manifestation of scattering is the red color of the sky during

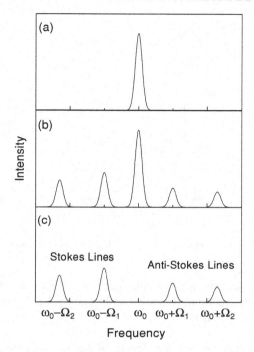

Figure 1.14   The spectral manifestation of the Raman effect. (a) The spectrum of the incident light. (b) The spectrum due to scattered (Rayleigh and Raman) light. (c) The Raman spectrum. The relative intensities of the incident, Rayleigh, and Raman lines are quite different in real cases (see the text).

day or night breaks, or the blue sky during the day. Both occur as a result of Rayleigh scattering of the sunlight due to molecules in the atmosphere. This type of scattering is an elastic photon process (the scattered photon energy is equal to the incident photon energy) and it is much more intense for higher frequencies.

Inelastic photon scattering processes are also possible. In 1928, the Indian scientist C. V. Raman (who won the Nobel Prize in 1930) demonstrated a type of inelastic scattering that had already been predicted by A. Smekal in 1923. This type of scattering gave rise to a new type of spectroscopy, *Raman spectroscopy*, in which the light is inelastically scattered by a substance. This effect is in some ways similar to the Compton effect, which occurs as a result of the inelastic scattering of electromagnetic radiation by free electrons.

Figure 1.14 shows how the Raman effect is manifested spectrally. When light (usually laser light) of frequency ω_0 comes in over the sample (Figure 1.14(a)), the output spectrum of scattered light consists of a predominant line of the same frequency ω_0 and much weaker ($\sim 1/1000$ of the main band) side bands at frequencies $\omega_o \pm \Omega_i$ (Figure 1.14(b)). The main line corresponds to Rayleigh scattered light, while the

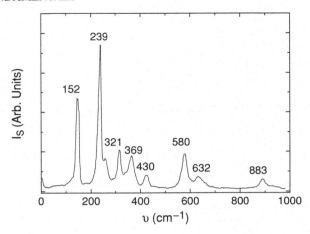

Figure 1.15   The Raman spectrum of lithium niobate at room temperature.

side-band spectrum (i.e., the spectrum of Figure 14(b), except for the main line at ω_0) is the actual Raman spectrum, as shown in Figure 1.14(c). This spectrum has the following particular properties:

(i) The Ω_i are the characteristic frequencies of the substance (in case of solids, they correspond to phonon frequencies).

(ii) Stokes and anti-Stokes lines (see Figure 1.14(c)) are always at frequencies located that are symmetrically to both sides of the main line (Rayleigh line) at ω_0.

(iii) The anti-Stokes lines are weaker than the Stokes lines.

(iv) The intensity of the lines is proportional to ω_0^4.

Raman spectra are usually represented by the intensity of Stokes lines versus the shifted frequencies Ω_i. Figure 1.15 shows, as an example, the Raman spectrum of a lithium niobate ($LiNbO_3$) crystal. The energies (given in wavenumber units, cm^{-1}) of the different phonons involved are indicated above the corresponding peaks. Particular emphasis will be given to those of higher energy, called *effective phonons* (883 cm^{-1} for lithium niobate), as they actively participate in the nonradiative de-excitation processes of trivalent rare earth ions in crystals (see Section 6.3).

Most of the four above-mentioned properties for Raman spectra can be explained by using a simple classical model. When the crystal is subjected to the oscillating electric field $\vec{E} = \vec{E}_0 e^{i\omega_0 t}$ of the incident electromagnetic radiation, it becomes polarized. In the linear approximation, the induced electric polarization in any specific direction is given by $P_j = \chi_{jk} E_k$, where χ_{jk} is the susceptibility tensor. As for other physical properties of the crystal, the susceptibility becomes altered because the atoms in the solid are vibrating periodically around equilibrium positions. Thus, for a particular vibrating mode (phonon) at frequency Ω, each component of the susceptibility tensor can be expressed as

$$\chi_{jk} = \chi_{jk}^{(0)} + \frac{\partial \chi_{jk}^{(0)}}{\partial Q} Q + \cdots \tag{1.22}$$

where $Q = Q_0 e^{\pm i\Omega t}$ represents a normal coordinate measured from the equilibrium position, indicated by the superscript (0) in the previous equation. Therefore, using Equation (1.22), the induced polarization can be written as

$$P_j = \chi_{jk} E_k = \chi_{jk}^{(0)} E_{0k} e^{i\omega_0 t} + Q_0 E_{0k} \frac{\partial \chi_{jk}^{(0)}}{\partial Q} e^{i\omega_s t} + \cdots \tag{1.23}$$

where we have denoted $\omega_s = \omega_0 \pm \Omega$. This expression corresponds to oscillating dipoles re-radiating light at frequencies of ω_0 (Rayleigh light), $\omega_0 - \Omega$ (Stokes Raman light), and $\omega_0 + \Omega$ (anti-Stokes Raman light). This explains the appearance of the Raman lines at symmetric frequencies in respect to ω_0, as stated in points (i) and (ii) above.

On the other hand, the radiated intensity of such oscillating dipoles is proportional to $\left| d^2 \vec{P}/dt^2 \right|^2$, so that we can write:

$$I \propto \omega_0^4 \left(\chi_{jk}^{(0)} E_{0k} \right)^2 + \omega_s^4 \left(Q_0 E_{0k} \frac{\partial \chi_{jk}^{(0)}}{\partial Q} \right)^2 + \cdots \tag{1.24}$$

The first term on the right-hand side of Equation (1.24) accounts for the generated intensity due to Rayleigh scattered light, while the second term is related to the intensity of the Raman scattered light. For visible light $\omega_0 \sim 10^{15}$ Hz, while the characteristic phonon frequencies are much shorter, typically $\Omega \sim 10^{12}$ Hz. Then $\omega_0^4 \approx \omega_s^4$, and the intensity of Raman scattering varies as ω_0^4, as stated in point (iv) above.

Property (iii) of the Raman effect can also be regarded from a quantum mechanical viewpoint by using the energy-level scheme displayed in Figure 1.16. In this quantum system, $\hbar\Omega$ corresponds to the energy of a real vibrational (phonon) state and the incident photon of energy $\hbar\omega_0$ is absorbed by exciting the system up to a virtual state. The Stokes Raman scattering occurs as a result of photon absorption from the ground state to a virtual state, followed by a depopulation to a phonon-excited state. On the other hand, the anti-Stokes Raman scattering is explained as being a result of photon absorption from the phonon-excited state to a virtual state, followed by a depopulation down to the ground state. Because of the Boltzmann population factor, $e^{-\hbar\Omega/kT}$, the phonon-excited state is less populated than the ground state and so the anti-Stokes lines must be of a lower intensity than the Stokes lines, as stated in point (iii) above.

It is important to recall that the virtual levels in Figure 1.16 do not correspond to real stationary eigenstates of our quantum system. As a result, Raman spectra are much weaker than fluorescence spectra (by an efficiency factor of about 10^{-5}–10^{-7}),

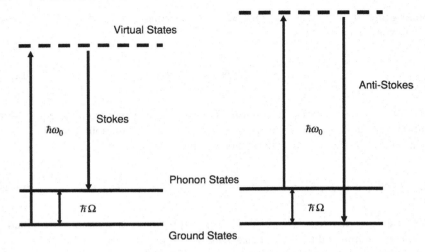

Figure 1.16   The energy-level scheme to account for Stokes and anti-Stokes Raman scattering.

as the latter make use of 'real electronic energy levels,' while 'virtual states' must be introduced to mediate in Raman spectra.

In *resonant Raman spectroscopy*, the frequency of the incident beam is resonant with the energy difference between two real electronic levels and so the efficiency can be enhanced by a factor of 10^6. However, to observe resonant Raman scattering it is necessary to prevent the possible overlap with the more efficient emission spectra. Thus, Raman experiments are usually realized under nonresonant illumination, so that the Raman spectrum cannot be masked by fluorescence.

The experimental arrangement for Raman spectroscopy is similar to that used for fluorescence experiments (see Figure 1.8), although excitation is always performed by laser sources and the detection system is more sophisticated in regard to both the spectral resolution (lager monochromators) and the detection limits (using photon counting techniques; see Section 3.5).

Raman spectroscopy is very useful in identifying vibration modes (phonons) in solids. This means that structural changes induced by external factors (such as pressure, temperature, magnetic fields, etc.) can be explored by Raman spectroscopy. It is also a very useful technique in chemistry, as it can be used to identify molecules and radicals. On many occasions, the Raman spectrum can be considered to be like a 'fingerprint' of a substance.

Finally, it should be mentioned that Raman and infrared absorption spectra (i.e., absorption spectra among vibrational levels) are very often complementary methods with which to investigate the energy-level structure associated with vibrations. If a vibration (phonon) causes a change in the dipolar moment of the system, which occurs when the symmetry of the charge density distribution is changed, then the vibration is *infrared active*. This means that $(\partial P/\partial Q)_0 \neq 0$. On the other hand, according to Equation (1.23), if a vibration causes a change in polarizability (or susceptibility), then $(\partial \chi/\partial Q)_0 \neq 0$, and so it is *Raman active*.

For local symmetries with a 'center of symmetry' (see Section 7.2), an infrared active vibration (phonon) is Raman inactive, and vice versa. This rule is usually known as the *mutual exclusion rule*.

1.6 ADVANCED TOPIC: THE FOURIER TRANSFORM SPECTROMETER

We have seen in the previous section that Raman spectra are complementary to infrared spectra. Both spectroscopies provide quite useful information on the phonon structure of solids. However, infrared spectra correspond to a range from about $100\ \text{cm}^{-1}$ to about $5000\ \text{cm}^{-1}$; that is, far away from the optical range. Thus, infrared absorption spectra are generally measured by so-called *Fourier Transform InfraRed (FTIR) spectrometers*. These spectrometers work in a quite different way to the absorption spectrophotometers discussed in Section 1.3.

The basic configuration of an FTIR spectrometer is schematically shown in Figure 1.17. The essential instrument of this spectrometer is a Michelson interferometer

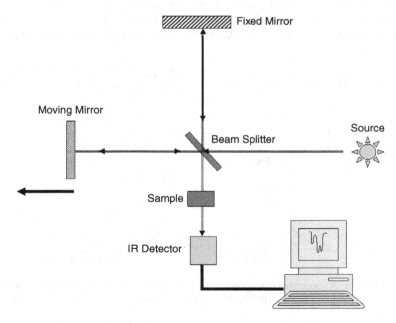

Figure 1.17   A schematic diagram of an FTIR spectrometer.

that consists of a fixed mirror, a moving mirror, and a beam splitter. When the IR beam, coming from a source, reaches the beam splitter, it is divided into two halves. Half of the beam is reflected in the beam splitter to the fixed mirror, while the remaining half passes through the beam splitter toward the moving mirror. These two beams are reflected in the fixed and moving mirrors, respectively, and come back to the beam splitter, where they recombine into a new beam that passes through the sample and is finally focused on to the detector.

Let us first consider an input monochromatic IR beam of wavelength λ, propagating in the empty interferometer (i.e., without a sample). If the two mirrors (moving and fixed) are at the same distance, L, from the beam splitter, the two light beams that recombine at the beam splitter travel the same distance $2L$ (i.e., there is no extra distance between them; $x = 0$) and they interfere constructively, so that a maximum is observed in the detector. If the moving mirror is displaced away from the beam splitter by a distance $\lambda/2$, the beam reflected in the moving mirror will now travel an extra distance of $x = 2(\lambda/2) = \lambda$, and constructive interference will again be observed in the detector. However, for a moving mirror displacement of $\lambda/4$, the beam reflected in the moving mirror will now travel an extra distance of $x = 2(\lambda/4) = \lambda/2$. Then, the two halves recombine in the beam splitter with a retardation of $\lambda/2$ to each other and so they interfere destructively, giving a minimum in the detector. In order to register a spectrum, the moving mirror is displaced along a distance $\gg \lambda$. Thus, for the case of a monochromatic beam of a given wavelength (or a given wavenumber), we should obtain an interference pattern like the one shown in Figure 1.18(a), with maximum detector signals for a mirror displacement that gives a retardation of $x = n\lambda$ and minimum detector signal for $x = (n + \frac{1}{2})\lambda$ ($n = 0, 1, 2, \ldots$). This interference pattern is called an *interferogram*. At this point, it should be noted that the Fourier

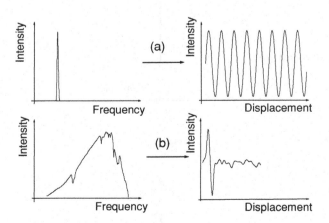

Figure 1.18   (a) A spectrum corresponding to a monochromatic IR beam (left) and its corresponding interferogram (right). (b) The broadband spectrum of a lamp (left) and its interferogram (right).

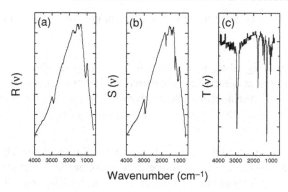

Figure 1.19 (a) A reference spectrum measured through an empty spectrometer. (b) A spectrum measured with an absorbing sample. (c) The final transmittance spectrum of the sample.

transform of the interferogram, $I(x)$, is just the spectrum $I(\nu)$ of the monochromatic radiation, shown on the left-hand side of Figure 1.18(a).

Let us now consider that an infrared lamp is used as the light source. In this case, a large number of wavelengths are emitted and the interferogram measured in the detector becomes much more complicated than for a monochromatic source. Figure 1.18(b) shows a typical shape of the spectrum of an IR lamp source and the corresponding interferogram obtained with the empty interferometer. For usual data acquisitions with an FTIR spectrometer, the interferogram $I(x)$ is measured and then it must be converted into a spectrum by means of a Fourier transformation. This transformation involves a certain amount of mathematical manipulation (using numerical methods) of the $I(x)$ signal, so that a good computer is one of the most important components of an FTIR spectrometer.

FTIR spectrometers generally provide transmittance spectra, $T(\nu)$. Three steps are realized before the final spectrum of a sample is obtained:

(i) An interferogram is measured with the empty spectrometer (without a sample). It is then Fourier transformed to give the *reference (lamp) spectrum $R(\nu)$*, like the one shown in Figure 1.19(a) for a real IR source.

(ii) Then, an interferogram with the sample inserted in the spectrophotometer is measured. After a Fourier transformation we obtain the so-called *sample spectrum $S(\nu)$* (Figure 1.19(b)). Notice that, as the sample is introduced, certain frequencies of the reference spectrum are selectively absorbed. Thus $S(\nu)$ looks very similar to $R(\nu)$, but with a lower intensity at those wavenumbers at which the sample absorbs.

(iii) Finally, the transmittance spectrum of the sample is obtained as the ratio of the sample and reference spectra, $T(\nu) = S(\nu)/R(\nu)$ (Figure 1.19(c)).

FTIR spectrometers have a number of advantages over conventional dispersive systems:

- They have a good signal-to-noise ratio, due to rapid data collection. This is because all of the data $S(\nu)$ are practically measured at the same time by the detector (the x scan is very fast), at variance with dispersive systems (such as the spectrophotometer discussed in Section 1.3), for which the frequency must be changed step by step (*the multiplex advantage*).

- They are especially attractive for analyzing weak light signals. This is because in an FTIR spectrometer all of the frequencies arrive at the detector simultaneously, so that the energy throughput is much greater than for dispersive instruments (the *throughput advantage*).

- Very accurate frequency determinations are possible, based on the use of the monochromatic light of a He–Ne laser for calibration (the *Connes' advantage*).

- FTIR spectrometers provide a high resolution compared to dispersive instruments. Moreover, this resolution is constant over the full spectral range.

Among the disadvantages of these spectrometers, we can find their usually high prices and the requirement of computational work to determine the spectra from the interferograms, which can introduce some artifacts.

EXERCISES

1.1. The Nd^{3+} ion shows several absorption lines of different widths in crystals. One of the absorption lines of Nd^{3+} ion in yttrium vanadate (YVO_4) peaks at 809 nm and presents a natural broadening at 2 K with a full width at half maximum of $(\Delta \nu)_{nat} = 18\,GHz$. Estimate (a) The natural broadening in the wavelengths units (nm) and (b) the lifetime of the excited state.

1.2. Two samples of the same phosphor crystal have quite different thicknesses, so that one of them has a peak optical density of 3 at a frequency of ν_0, while the other one has a peak optical density of 0.2 at ν_0. Assume a half width at half maximum of $\Delta \nu = 1\,GHz$ and a peak wavelength of 600 nm, and draw the absorption spectra (optical density versus frequency) for both samples. Then show the absorbance and transmittance spectra that you expect to obtain for both samples and compare them with the corresponding absorption spectra. (To be more precise, you can suppose that both bands have a Lorentzian profile, and use expression (1.8), or a Gaussian line shape, and then use expression (1.9).)

1.3. An optical material has an absorption coefficient of $10\,cm^{-1}$ at 400 nm. (a) What thickness of material must you cut to fabricate an optical filter with optical density of 3 at 400 nm? (b) If you have a laser of 1 W power at this wavelength but you want to illuminate a sample with only 1 m W, how many of these filters do you need?

1.4. An optical filter has optical density of 4.0 at a wavelength of 633 nm. (a) Determine the transmittance and the absorbance of the filter at this wavelength. (b) If a laser beam of 1 mW power at 633 nm passes through such a filter, determine the laser power beyond the filter.

1.5. The quality of the light transmission of an optical fiber is usually given by its *attenuation*. The attenuation, At, of a beam of incident intensity I_0 after traversing a length l through a fiber is defined by $At(\text{dB}) = 10 \log [I_0/I(l)]$, where $I(l)$ is the light intensity at l and dB indicates decibels units. This length is usually taken as $l = 1$ km, so each fiber is characterized by its attenuation in dB km^{-1}. (a) Assuming that the attenuation is only due to absorption (normally a fraction of light is also lost by scattering), establish the relationship between the attenuation, dB km^{-1}, and the absorption coefficient α of the fiber. (b) Considering a fiber with a typical attenuation coefficient of 0.2 dB km^{-1}, estimate its absorption coefficient. (c) Estimate the optical density that would be measured for 1 km of fiber. (d) If you want to transmit information with a laser of 1 W, estimate the laser power after 50 km of fiber.

1.6. A crystal activated with Ti^{3+} ions presents an absorption that peaks at 514 nm and the corresponding emission spectrum peaks at 600 nm. A sample of this crystal, which has an optical density of 0.6 at the absorption peak, is illuminated with an Ar laser emitting at 514 nm with a power of 2 mW. (a) Determine the laser power of the beam after it passes through the crystal. (b) If the quantum efficiency is $\eta = 0.6$, determine the intensity (in photons per second) emitted as luminescence and the power dissipated as heat in the crystal.

1.7. The yttrium aluminum garnet crystal, $Y_3Al_5O_{12}$, doped with Nd^{3+} ions, is a well-known solid state laser material (abbreviated to Nd:YAG). If the fluorescence lifetime of the main laser emission is 230 μs and the quantum efficiency of the corresponding emitting level is 0.9, determine (a) the radiative lifetime and (b) the radiative and nonradiative rates.

1.8. A phosphor emits a weak light at 400 nm when illuminated with infrared light of wavelength 1 μm. (a) Make an energy diagram to explain this luminescence. (c) How many photons must be sequentially absorbed per center in order to observe this luminescence? (c) Determine the dependence that you expect of the emission intensity at 400 nm on the excitation intensity at 1 μm.

1.9. Laser light of 514 nm illuminates a crystal of $LiYF_4$ and scattered lines are detected at 514 nm, 518.1 nm, 518.7 nm, 519.3 nm, 520.6 nm, 521.1 nm, 522.8 nm, and 525.5 nm. (a) Estimate the energies of the phonons responsible for the Raman spectra. (b) Determine the peak positions of the anti-Stokes lines. (c) Assuming that the intensity of the Raman peak at 518.1 nm is I_1 at 300 k, make a rough estimation of the intensity for the corresponding anti-Stokes line.

REFERENCES AND FURTHER READING

Fox, M., *Optical Properties of Solids,* Oxford University Press, Oxford (2001).

Gaft, M., Reisfeld, R., Panczer, G., Ioffe, O., and Sigal, I., *J. Alloys Compounds,* **323–324**, 842–846 (2001).

Lifante, G., Jaque, F., Hoyos, M. A., and Leguey, S., *J. Gemmology*, **3**, 22 (1990).

Jaque, D., Capmany, J., Rams, J., and García Solé, J., *J. Appl. Phys.*, **87**(3), 1042–1048 (2000).

Roland Menzel, E., *Laser Spectroscopy. Techniques and Applications*, Marcel Dekker, Inc., New York (1995).

Saleh, B. E. A. and Teich, M. C., *Fundamentals of Photonics*, John Wiley & Sons, Inc., New York (1991).

Svelto, O., *Principles of Lasers*, Plenum Press, New York (1986).

2
Light Sources

2.1 INTRODUCTION

Among other modern and interesting applications, light has proven to be a very useful tool in measuring and scanning techniques. This is particularly true in the field of optical spectroscopy, where the analysis of light–matter interaction phenomena provides fundamental information about the nature of both matter and light. Here, light must be understood broadly, including radiation within the 'optical range' as defined in Section 1.2. In this chapter, we will devote our attention to describing the physical basis of different types of light sources as fundamental instruments in optical spectroscopy.

Due to its relevance in the operational mechanism of various lamps, let us begin with a brief description of the main features of thermal radiation.

2.1.1 Thermal Radiation and Planck's Law

Let us consider *thermal radiation* in a certain cavity at a temperature T. By the term 'thermal radiation' we mean that the radiation field is in thermal equilibrium with its surroundings, the power absorbed by the cavity walls, $P_a(\nu)$, being equal to the emitted power, $P_e(\nu)$, for all the frequencies ν. Under this condition, the superposition of the different electromagnetic waves in the cavity results in standing waves, as required by the stationary radiation field configuration. These standing waves are called *cavity modes*.

By means of geometric arguments, it can be shown that the number of modes $n(\nu)$ per unit volume within a frequency interval $d(\nu)$ is given by

$$n(\nu)d\nu = \frac{8\pi \nu^2}{c^3}d\nu \tag{2.1}$$

where c is the speed of light in the vacuum.

An Introduction to the Optical Spectroscopy of Inorganic Solids   J. García Solé, L. E. Bausá, and D. Jaque
© 2005 John Wiley & Sons, Ltd   ISBNs: 0-470-86885-6 (HB); 0-470-86886-4 (PB)

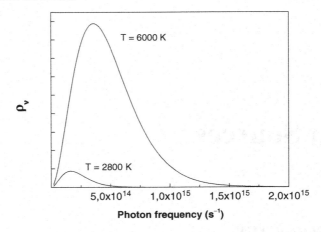

Figure 2.1   The spectral distribution of the energy density for two different temperatures.

As Planck suggested in 1900, each mode of the radiation field can only emit or absorb energy in discrete amounts that are integer multiples of a minimum energy quantum $h\nu$. Thus, taking into account expression (2.1) and the fact that in thermal equilibrium the distribution of the total energy into the different cavity modes is governed by the Maxwell–Boltzmann distribution, we can obtain the *energy density*, $\rho_\nu d\nu$, of the radiation field within a frequency interval ranging from ν to $\nu + d\nu$, as

$$\rho_\nu d\nu = \frac{8\pi \nu^2}{c^3} \frac{h\nu}{e^{h\nu/\kappa T} - 1} d\nu \tag{2.2}$$

This is Planck's famous radiation law, which predicts a *spectral energy density, ρ_ν*, of the thermal radiation that is fully consistent with the experiments. Figure 2.1 shows the spectral distribution of the energy density ρ_ν for two different temperatures. As deduced from Equation (2.2), the thermal radiation (also called *blackbody radiation*) from different bodies at a given temperature shows the same spectral shape. In expression (2.2), ρ_ν represents the energy per unit time per unit area per frequency interval emitted from a blackbody at temperature T. Upon integration over all frequencies, the total energy flux (in units of W m^{-2}) – that is, $E_{tot} = \int_0^\infty \rho_\nu \, d\nu$ – yields

$$E_{tot} = \sigma T^4 \tag{2.3}$$

which is known as the *Stefan–Boltzmann law*. Here, $\sigma = 5.67 \times 10^{-8}$ W m^{-2} K^{-4} is the Stefan–Boltzmann constant. In real cases, the blackbody radiation actually obeys the slightly modified form of Equation (2.3) given by

$$E_{tot} = \varepsilon \sigma T^4 \tag{2.4}$$

where the coefficient ε is the *emissivity* of the surface, which corresponds to the fraction of the ideal blackbody spectrum energy that a real body actually emits $(\varepsilon = 1$ in the ideal case).

An important feature of the thermal radiation field, described by its energy density ρ_ν, is that it is isotropic; that is, the emission is the same in all directions.

Some important examples of real radiation sources with spectral distributions close to the Planck distribution are the sun (which shows a spectrum consistent with $T = 6000$ K) and the bright tungsten wire of a light bulb ($T = 2800$ K).

On the other hand, examples of nonthermal isotropic radiation sources are spectral lamps, which emit discrete spectra.

Finally, lasers are examples of nonthermal and anisotropic radiation sources. The radiation field is concentrated in a few modes. Moreover, this radiation is emitted into a small solid angle.

EXAMPLE 2.1   *Consider the thermal radiation field at room temperature (300 K). Determine the number of modes per m^3 and the average number of photons per mode in the visible range for the spectral interval $d\nu = 10^9\ s^{-1}$.*

We chose a wavelength of $\lambda = 600$ nm in the visible spectrum. For that wavelength, the corresponding frequency is $\nu = (c/\lambda) = 5 \times 10^{14}\ s^{-1}$.

According to Equation (2.1), the mode density at that frequency in a spectral interval of $10^9\ s^{-1}$ is

$$n(\nu)d\nu = (2.3 \times 10^5) \times 10^9 = 2.3 \times 10^{14}\ m^{-3}$$

This value is some orders of magnitude higher than the mode density at lower frequencies, such as is the case of microwaves, but smaller than that at higher frequencies, such as in the case of X-rays.

To determine the average number of photons per mode, we should first determine the mean energy, W, for the particular mode at $\nu = 5 \times 10^{14}\ s^{-1}$ at 300 K. According to the Maxwell–Boltzmann distribution, this is given by

$$W = \frac{h\nu}{e^{h\nu/\kappa T} - 1} = 5.56 \times 10^{-54}\ J$$

Since the energy of a photon is $h\nu = 3.3 \times 10^{-19}$ J, the average number of photons for that mode is 1.7×10^{-35}. This indicates that the number of photons per mode in the visible region for a spectral width of $10^9\ s^{-1}$ (in the order of the Doppler width) is a very small number compared to unity. This has important consequences, as we will see in Section 2.3.

2.2 LAMPS

Different types of lamps have been used in numerous applications, including spectroscopy, material analysis, and laser pumping. We will enumerate the main characteristics of some of the most frequently used systems.

2.2.1 Tungsten and Quartz Halogen Lamps

Tungsten lamps are used in spectrometers since they can supply a useful spectrum in the near infrared region. In these lamps, electricity heats a coil or tungsten wire hot enough to make it glow, in a process called incandescence. A usual temperature is 2800 K, for which we get a very bright yellow (nearly white) color. That is the temperature of a normal light bulb filament. This filament is inside a glass envelope that contains argon or nitrogen gas. The spectral density is represented in Figure 2.1.

As a variation on these bulb lamps, we should mention halogen lamps. These also use a tungsten filament, but it is encased inside a much smaller quartz envelope. Because the envelope is close to the filament, it would melt if it were made from glass. The gas inside the envelope is also different. It consists of a gas from the halogen group. These gases have a very interesting property: if the temperature is high enough, the halogen gas will combine with tungsten atoms as they evaporate and redeposit them on the filament. This recycling process allows the filament to last a lot longer. In addition, it is now possible to run the filament at a higher temperature, which implies a greater light intensity per unit of energy.

EXAMPLE 2.2   *Estimate the total power radiated by a surface of 0.1 cm² at a temperature of 2000 K, assuming that the emissivity of this surface is 0.8.*

According to Equation (2.4), $E_{tot} = \varepsilon \sigma T^4 = 0.8 \times 5.67 \times 10^{-8} \times (2000)^4$ W m^{-2} $= 7.25 \times 10^5$ W m^{-2}. Therefore, the total power radiated by the surface of area 0.1 cm^2 is $P_{tot} = 7.25$ W.

2.2.2 Spectral Lamps

Spectral lamps find a home in special laboratory applications, where they are commonly employed as stable sources of discrete spectral lines, which correspond to the atomic spectra of specific elements (metals, in most cases). Originally, the atomic spectra were produced either by creating an arc between electrodes fabricated from the emitting metal within a discharge tube, or by sprinkling a powdered salt into an ordinary gas flame. Nowadays, electrical discharge lamps have been developed with very good performance, due to the high purity of the metals contained within the discharge tube. In the category of spectral lamps we also include those gas discharge lamps that contain gases such as Ne, Xe, or He at a low–medium pressure (less than 1 atmosphere). The atoms of the gas are excited to high energy levels by an electric current flowing through the gas. Figures 2.2(a) and 2.2(b) show, as examples, the spectra of low-pressure sodium and mercury vapor discharge lamps. As shown, nearly monochromatic resonance lines are observed due to the transitions among the different energy levels of the vapor metal.

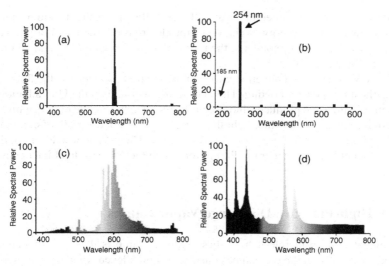

Figure 2.2 The emission spectra of different spectral lamps: (a) a low-pressure sodium lamp; (b) a low-pressure mercury lamp; (c) a high-pressure sodium lamp; (d) a high-pressure mercury lamp.

2.2.3 Fluorescent Lamps

Fluorescent lamps are based on low-pressure gas (mercury, in most cases) discharge lamps. The central element in a fluorescent lamp is a sealed glass tube, as shown in Figure 2.3. The tube contains a small amount of mercury and an inert gas, which are kept under very low pressure (a few hundredths of an atmosphere). The tube also contains a phosphor powder, which coats the inside of the glass. The tube has two electrodes, one at each end, connected to an electrical circuit. When the lamp is turned on, electrons from the electrodes migrate through the gas from one end of the tube to the other. As electrons and charged atoms move through the tube, some of them will

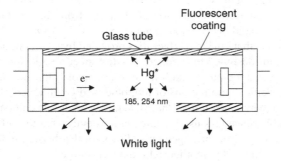

Figure 2.3 The schematic design of a fluorescent lamp.

collide with the gaseous mercury atoms. These collisions excite the atoms, pumping electrons up to higher energy levels. When the electrons return to their original energy level, they release light photons in the ultraviolet (UV) region (185 and 254 nm), as shown in Figure 2.2(b).

This UV radiation is converted into visible radiation by means of the fluorescence of the phosphor powder coating. The phosphor material uses the UV radiation as an excitation source and produces fluorescent emission in the visible region, with a broad spectrum to give off the white light that we can see. A good variety of combinations of phosphors are used (Shionoya and Yen, 1999). The principal field of application of fluorescent lamps is general lighting, for which they constitute efficient devices.

2.2.4 High-Pressure Discharge Vapor Lamps

The fundamental principle behind these types of lamp is that the monochromatic resonance lines from the gas can be considerably broadened when the gas pressure is increased. A collision-induced additional broadening (often called pressure broadening) affects the energy levels of the atomic gas. This effect is due to the different shifts of the energy levels produced by the mutual interactions of atoms at short distances. Instead of the the typical lines of a diluted gas, the discharge through the high-pressure gas produces emission bands. Figures 2.2(c) and 2.2(d) show the spectra of a high-pressure discharge sodium and a mercury lamp, compared to those of low-pressure lamps. Typical pressures in these lamps are higher than 200 atmospheres.

These lamps show moderately good color rendering properties, the very broad spectrum being one of their main utilities. For instance, high-pressure Xe lamps or deuterium lamps are commonly used in optical spectroscopy techniques.

2.2.5 Solid State Lamps

Finally, we should mention the so-called solid state lamps, which are based on semiconductor technology, namely the *light emitting diodes* (LED). Here, the origin of the radiation is the radiative recombination of electrons from the conduction band with holes in the valence band in a p–n junction (see Section 2.4). One of the most interesting characteristics is that, due to the facilities provided by semiconductor technology, they can be custom-designed to produce radiation over a wide spectral range, from the UV to the infrared region. An additional selection of colors is also made possible by controlling specific combinations of red, blue, and green LEDs. The applications mainly include displays and outdoor lamps. White sources that could replace tungsten lamps are available nowadays using source color devices based on GaN and InGaN diodes. The flexibility of the arrangements allows for the thermal radiation spectra to be artificially reproduced by nonthermal radiation sources.

2.3 THE LASER

2.3.1 Lasers as Light Sources in Spectroscopy

It is already a fact that lasers are replacing conventional lamps in a great variety of spectroscopic applications. The origin of this substitution lies in their superior performance over incoherent light in many experimental situations. Many spectroscopic experiments have been improved, and moreover new techniques have been developed due to the particular advantages provided by lasers. The characteristics of laser radiation on their own constitute real advantages and justify their widespread use in many applications.

Let us mention some of those characteristic features and give some examples of their influence in the field of spectroscopy:

(i) First, there is the large *spectral density of power* attainable from most types of lasers, which can exceed that from incoherent light sources by several orders of magnitude. This feature has many consequences. On one hand, noise problems (either coming from background radiation or due to the detector) can be significantly reduced, which implies an improvement of the signal-to-noise ratio. On the other hand, the high intensity of the radiation provided by lasers allows the study of nonlinear phenomena, such as multiphoton processes or saturation phenomena, which otherwise were not easily accessible by conventional linear spectroscopy.

(ii) Another important advantage of lasers over conventional lamps is the *small divergence of the radiation beam*. This makes the handling and control of the light beam much easier and allows its confinement in integrated optics devices. From the viewpoint of spectroscopy, some examples of the advantage of a small divergence in collimated laser beams can be mentioned: measurements of very small absorption coefficients, which can be achieved by using long path lengths through the sample (even by using a wave-guided configuration); and the efficient collection and imaging onto the detection system of the incoming radiation from a very small interaction zone of the sample. This can be particularly useful, for instance, in Raman scattering or low-level fluorescence spectroscopy.

(iii) The *narrow spectral width* that can be obtained with some type of lasers constitutes an advantage that has a particular impact on the development and application of high-resolution spectroscopy techniques. In fact, the spectral resolution provided by some lasers may exceed that of the largest monochromators by several orders of magnitude.

(iv) The possibility of continuously *tuning* the wavelength, with all of the aforementioned characteristics. Actually, the use of lasers in spectroscopy can turn out to be less expensive than it might seen, since a *tunable laser* could replace an intense source of radiation and a high-resolution, high-priced monochromator.

(v) The possibility of pulsed lasers supplying intense short and *ultra-short pulses* up to the femtosecond range. This has permitted the study of fast and ultrafast phenomena, such as various fast relaxation processes in solids.

(vi) With laser beams, we can deal with the phenomena that occur when many mutually coherent waves are superimposed. Many experiments in laser spectroscopy depend on the *coherence* properties of the radiation. From the viewpoint of spectroscopy, the use of coherent light beams can be applied to measurements that require a high resolving power. This is the case in interference spectroscopy. For instance, by using Fabry–Perot resonators or difference-frequency analysis methods, the superposition of beams with slightly different frequencies leads to beats, and thus to a time-dependent intensity that can be followed with sufficiently fast photodiodes. The spectral shape of those beams can be then resolved (Lauterborn and Kurz, 2003).

Laser radiation can be obtained nowadays over a wide spectral range from the ultraviolet to the far infrared region, covering the range of optical spectroscopy. Figure 2.4 shows schematically the spectral zones covered by different types of lasers. Although there are some specific regions in which direct laser action is not available,

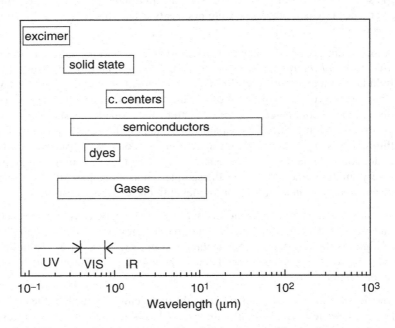

Figure 2.4   The spectral regions covered by different types of lasers ('c. centers' stands for color center lasers).

coherent radiation can be achieved at those wavelengths by frequency-mixing techniques from laser beams emitting at other wavelengths (see Section 2.5).

2.3.2 The Basic Principles of Lasers

The word *LASER* is an acronym for *Light Amplification by Stimulated Emission of Radiation*. Similar to the way in which transistor systems are available to generate and amplify electrical signals, with the advent of lasers we have at our disposal devices that are able to generate and amplify coherent light.

The essential elements of a laser device are as follows:

• An *active medium*, consisting of a collection of atoms, molecules, or ions in a gaseous, liquid, or solid state, which generates and amplifies light by means of appropriate transitions between its quantum energy levels.

• A *pumping* process, to excite those atoms (molecules, ions, etc.) up to higher quantum energy levels to produce *population inversion*.

• An *optical resonator* system, which provides the *optical feedback*.

Both the active medium and the resonator determine the light frequencies generated.

The light–matter interaction process that takes place in the active medium constitutes the essential key of laser radiation. On the basis of phenomenological considerations, Einstein developed a theory that permits a qualitative understanding of the processes related to light absorption and emission by atoms. Three basic light–matter interaction processes can be considered: *absorption, spontaneous emission*, and *stimulated emission* of photons.[1] Figure 2.5 illustrates those processes for a two energy level system (E_1 and E_2). The basis of laser radiation lies in the last process: stimulated emission. That is, when the atomic system has absorbed a photon of energy $h\nu = E_2 - E_1$, and thus the upper level is populated, a second photon of the same energy $h\nu$ may cause this energy to be emitted as a photon. Then, two photons with identical properties, one incident and the other emitted due to stimulation leave the system. In terms of waves, we can say that the electromagnetic wave associated with the stimulated emission has the same direction, phase, polarization, and wavelength as the incident radiation. Thus, the stimulated emission process in itself constitutes an optical amplification phenomenon, since the energy liberated due to the stimulated transition adds to the incident wave on a constructive basis, reinforcing the light beam.

The quantum mechanical perturbation theory that allows us to calculate the probability of absorption and stimulated emission shows (see Section 5.3) that both

---

[1] The cover of this book shows each of these three processes: the stimulated emission represented by the blue light of an Ar^+ laser; the absorption process responsible for the attenuation of the blue laser in a $LiNbO_3:Pr^{3+}$ crystal; and the spontaneous emission that corresponds to the red light emitted from Pr^{3+} ions.

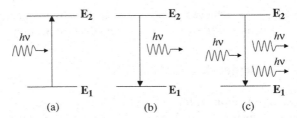

Figure 2.5 Descriptions of different light–matter interaction processes: (a) absorption; (b) spontaneous emission; (c) stimulated emission.

processes differ only in the initial conditions. Upon absorption, the system starts from a state of lower energy, whereas upon stimulated emission it starts from a state of higher energy. Hence, the transition probability is equal for both processes, provided that the levels have equal degeneracy.

2.3.3 Population Inversion: the Threshold Condition

It is clear that in order to get stimulated emission, a pumping process is required to excite the system into its high quantum energy level. Real materials can be pumped in many ways, as will be mentioned later. For laser action to occur, the pumping process must produce not merely excited atoms, but the condition of *population inversion*.

Let us assume an active medium that responds to the energy-level diagram of Figure 2.6(a). It consists into four energy levels E_i with respective population densities N_i ($i = 0, \ldots, 3$). Let us also assume that laser action can take place due to the stimulated emission process $E_2 \rightarrow E_1$. When a monochromatic electromagnetic wave with frequency v, such as $(E_2 - E_1)/h = v$, travels in the z direction through the medium, the intensity of the beam at a depth z into the crystal is given by

$$I(v, z) = I_0 e^{\sigma(N_2 - N_1)z} \qquad (2.5)$$

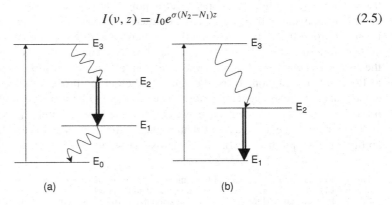

Figure 2.6 An energy-level scheme for (a) four- and (b) three-level lasers. $\rightarrow$, Pumping transition; $\Rightarrow$, laser transitions; $\text{\tiny WW}\!\!\blacktriangleright$, fast nonradiative transitions.

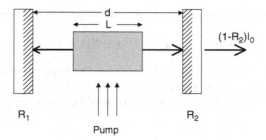

Figure 2.7   A schematic design of a laser.

where I_0 is the intensity of the incident beam, and σ is the cross section of the transition, defined in Equation (1.5). In the former expression the sign of the exponent, specifically that of the population difference between the involved levels, is a key factor to get amplification from the medium. For $N_2 > N_1$, the incident beam can be amplified and optical gain is possible from the system; otherwise, the beam will be attenuated as explained in Section 1.3. The condition $\Delta N = N_2 - N_1 > 0$ is known as *population inversion* with respect to the thermal equilibrium situation in which, according to the Boltzmann distribution, the population distribution for the energy levels E_2 and E_1 requires that $N_2 < N_1$. As will be seen, a threshold condition is needed for the population inversion $\Delta N = N_2 - N_1$.

Expression (2.5) is, in fact, the basis on which the gain from the active medium is defined. If the active medium is placed between two mirrors, as shown in Figure 2.7, the wave is reflected back and forth, traversing the amplifying medium many times, and increasing the total amplification. Assuming a system without losses, for a given length L of the active medium the *gain per round trip*, $G(v)$, is defined by

$$I(v, 2L)/I(v, 0) = e^{G(v)} \tag{2.6}$$

so that, according to Equation (2.5),

$$G(v) = 2\sigma(N_2 - N_1)L \tag{2.7}$$

However, a series of factors introduce *losses* in the system; namely, the reflectivities of the mirrors (R_1 and R_2) on the figure, which reflect only a fraction, R_1 and R_2, of the intensity. Additional losses can be produced by absorption in the windows of the cell that contains the active medium (if this is the case), diffraction by apertures, and scattering due to particles or imperfect surfaces. All of those losses can be included in a *loss factor* per trip, expressed as $e^{-\gamma}$. Thus, considering both amplification and intensity decrease per round trip, the intensity after a single round trip through a resonator of length d is

$$I(v, 2d) = I(v, 0)e^{(G - \gamma)} \tag{2.8}$$

The wave is amplified if the gain overcomes the losses per round trip; that is,

$$G > \gamma \quad \text{or} \quad (N_2 - N_1) > \gamma/2\sigma L \tag{2.9}$$

This implies a threshold condition $(G = \gamma)$ for the population difference $\Delta N_{th} = (N_2 - N_1)_{th} = \gamma/2\sigma L$. If the population inversion, ΔN, is larger than ΔN_{th}, a wave reflected back and forth in the resonator will be amplified and its intensity will increase, in spite of the losses.

EXAMPLE 2.3   *Suppose that a 7.5 cm length rod of a $Nd^{3+} : Y_3Al_5O_{12}$ (Nd:YAG) laser crystal is located in a linear cavity with two mirrors of transmittances $T_1 = 0$ and $T_2 = 0.5$ at the laser wavelength (1.06 μm). If the cross section is $\sigma = 8.8 \times 10^{-19} cm^2$, determine the population inversion density at threshold. Assume that losses are only due to the output mirror transmittance.*

With the reflectivities $R_1 = 1 - T_1$ and $R_2 = 1 - T_2$ of the resonator mirrors, the intensity I of a wave in a passive resonator after a single round trip is given by $I = R_1 R_2 I_0 = I_0 \exp(-\gamma_R)$, γ_R being the reflection losses and I_0 the incident intensity. Taking into account that the losses are only due to the output coupler (that is, $\gamma = \gamma_R$) and that the reflectivity of the output mirror is given by $R_2 = 1 - T_2 = 0.5$, the population inversion density at threshold is given by

$$(N_2 - N_1)_{th} = -\frac{\ln(R_1 R_2)}{2\sigma L} = -\frac{\ln(0.5)}{2 \times 8.8 \times 10^{-19} cm^2 \times 7.5\,cm}$$
$$= 5.25 \times 10^{16} cm^{-3}$$

where we have assumed that the input and output couplers are coated at both of the end faces of the laser crystal (the resonator length coincides with the crystal length).

A usual concentration of Nd^{3+} ions for laser applications is $6 \times 10^{19} cm^{-3}$. In such a case, the population inversion density at threshold represents a fraction of about 1000 ppm of the total concentration.

In a laser system, the wave is initiated by spontaneous emission from the excited state atoms in the active medium. The spontaneously emitted photons traveling parallel to the resonator axis are able to create new photons by stimulated emission. Above the threshold they induce a photon avalanche, which grows until the depletion of the population inversion compensates the repopulation due to pumping.

The characteristics of the active medium determine those of the laser action. According to Equation (2.7), the gain $G(\nu)$ is directly related with the particular characteristics of the quantum energy levels of the active medium via the transition cross section σ. As follows from Sections 1.3 and 1.4 (see also Exercise 5.4):

$$\sigma = \left(\frac{\lambda}{2}\right)^2 \frac{g(\Delta\omega)}{\tau_0} \tag{2.10}$$

where λ is the emission wavelength, $g(\Delta\omega)$ is the line profile, and τ_0 is the spontaneous emission lifetime of the upper laser level. These magnitudes are determined by the energy levels involved, which depend on the nature of the active medium.

The active medium also determines the pumping scheme. Commonly, two types of operational schemes are used to describe laser operation: *four-level* and *three-level* laser systems:

(i) *The four-level operational scheme.* Figure 2.6(a) shows a simplified scheme for four-level laser operation: many real laser systems are based on this operational scheme. Atoms are pumped by some pumping mechanism from the ground level E_0 into some excited level E_3. Then, they relax down into the *upper laser level* E_2 from which stimulated $E_2 \rightarrow E_1$ transitions can occur. The basic requirement for obtaining population inversion is that atoms should relax from the *lower laser level* E_1 down to the fundamental level faster than the atoms relax from level E_2. Then, besides a strong pumping rate, the population inversion depends on the relaxation rates A_1 and A_2 of the lower and upper laser levels, respectively. The essential condition for population inversion is given by

$$A_1 > A_2 \tag{2.11}$$

In many real lasers, the upper laser level, E_2 in Figure 2.6(a), is a metastable level; that is, it has a long lifetime compared to the lower laser level ($\tau \gg \tau_1$). If we can pump efficiently into such a longer-lived upper level, and provided that there is a lower energy level, E_1 in Figure 2.6(a), with a short lifetime, then a population inversion is very likely to be established.

(ii) *Three-level operation.* Figure 2.6(b) illustrates the basis of a three-level laser operational scheme. It differs from the four-level case in that the lower laser level is the ground level E_1. This is a serious disadvantage, since more than half of the atoms initially in the ground state must be pumped through the pumping level E_3 into the upper laser level E_2 to achieve population inversion. Three-level lasers are therefore usually not as efficient as four-level lasers.

2.3.4 Pumping Techniques

In general, different pumping techniques are used in practical devices depending on the type of active medium. Gas discharges are among the most widely used pumping processes when using a gas laser medium. Direct electron impact with atoms or ions and transfer of energy by collisions between different atoms are the two main mechanisms involved.

Optical pumping techniques are also very common. The source of the pumping light may be a continuous arclamp, a pulsed flashlamp, another laser, or even focused sunlight.

Other more exotic mechanisms include chemical reactions in gases, high-voltage electron beam pumping of gases, and direct current injection across the junction of a semiconductor laser.

2.3.5 The Resonator

When a radiation source is placed inside a closed cavity, its radiation energy is distributed among all of the modes following Equations (2.1) and (2.2), once the system has reached equilibrium. As we have seen in Example 2.1, in spite of the large number of modes in such a closed cavity, the mean number of photons per mode corresponding to the optical region is very small. Specifically, it is very small compared to unity. This is the ultimate reason why, in thermal radiation fields, the spontaneous emission per mode by far exceeds the stimulated emission. (Remember that the stimulated emission process requires the presence of photons to induce the transition, opposite to the case of the spontaneous emission process.)

However, it is possible to concentrate most of the radiation onto a few modes in such a way that the number of photons in those modes becomes large and the stimulated emission in those modes will dominate (although the total spontaneous emission rate into all modes may still be larger than the induced rate in these few modes). Such selection of few modes is realized in a laser by using an appropriate resonator, which should exhibit a strong feedback for those modes. The resonator will allow an intense radiation field to be built in the modes with low losses, and will prevent oscillation from being reached in the modes with high losses.

Such a resonator can be realized with an *open cavity* consisting of two plane or curved mirrors, as represented in Figure 2.7 (*linear cavity*). Details of the stability conditions for different types of open resonators can be found elsewhere (Siegman, 1986). Other more sophisticated configurations, such as those of *ring cavity lasers* (Demtröder, 2003) and *microlasers* (Kasap, 2001) are also used.

Figure 2.8 sketches the *gain profile, $G(\nu)$,* of an active medium, together with the resonator modes and those modes for which the gain exceeds the losses. *Multimode* and *single-mode* operation will be obtained in practical systems depending on the oscillation thresholds for the different modes. If the threshold level is as in Figure 2.8(a), then several modes will contribute to the laser output. On the contrary, if threshold level is as high as that of Figure 2.8(b), the laser will operate in a single mode of frequency ν_0. The different geometric factors of the cavity (cavity length, intracavity optical elements, etc.) will determine the mode distribution of the laser output beam. A good description of different methods for obtaining single-mode lasers can be found in (Demtröder, 2003).

2.4 TYPES OF LASERS

Due to the large variety of laser materials and pumping methods, it is almost impossible to catalog all the laser devices that have been demonstrated up to date. However, we can make a classification of the laser systems based on the different types of active media. We will briefly comment on the basis and properties of some specific types of laser systems, which are representative of different laser schemes.

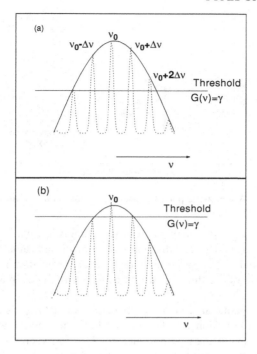

Figure 2.8 A schematic diagram of the gain spectral profile, $G(\nu)$, of a laser transition (solid line), together with the axial resonator modes (dotted line) of a cavity in which the frequency separation between adjacent modes is $\Delta \nu$. (a) Multimode and (b) single-mode operation. The frequencies of those modes for which the gain exceeds the losses have been marked.

2.4.1 The Excimer Laser

In this type of Laser, the active medium consists of an inert gas (X) or of a mixture of an inert gas and a halide gas (X + Y). The term *excimer* stands for 'excited dimmer' which refers to a diatomic molecule of two inert gas atoms (XX)* or a molecule of an inert gas atom and a halide gas atom (XY)*.

The determining feature by which laser action can be efficiently obtained from this type of active medium is the fact that the atoms that form the dimmer are only bound in the excited state. Figure 2.9 shows a schematic diagram of the laser energy levels in a molecule of excimer. The laser transition is produced between two molecular electronic levels in which the potential energy curve for the fundamental state is repulsive. This ensures the population inversion.

The chemical mechanisms required to achieve the upper laser state are quite complex, involving in some cases up to 82 different reactions. The promotion of molecules to the upper laser state requires reactions, which include ionization, dissociation, and

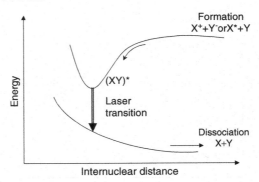

Figure 2.9   A schematic potential energy diagram of an excimer molecule.

atomic and molecular excitation processes. Accordingly, the pumping method should be carried out via a fast electrical discharge or by electron beams.

On the other hand, the energy released by the laser transition is associated with a dissociation process of the dimmer, which consequently takes place in the UV part of the spectrum.

Excimer lasers operate at typical powers of 50–200 mJ in pulses of around 20 ns. Table 2.1 shows the emission wavelengths for different excimer lasers.

Excimer lasers are of great importance for UV and vacuum UV (VUV) spectroscopy and photochemistry. They are also found in a wide range of applications. For example, they are used in micromachine medical devices, including refractive surgery, in photo-lithography for the microelectronics industry, for material processing, as optical pump sources for other type of lasers (dyes), and so on. More details about excimer lasers can be found in Rodhes (1979).

Table 2.1   Different types of excimer lasers and their emission wavelengths

Excimer	Operational wavelength (nm)
Ar_2	126
ArF	193
ArCl	175
Kr_2	146
KrF	248
KrCl	222
Xe_2	170–175
XeF	351, 353
XeCl	308
XeBr	282

2.4.2 Gas Lasers

Lasers belonging to this category can be classified according to different criteria. In general, we will distinguish two types of gas lasers, depending on the nature of their energy levels, and therefore, on the optical transitions participating in the laser action.

(a) The laser transition takes place between electronic quantum levels of neutral or ionized atoms in a gas. In particular, the laser transition occurs between electronic levels involving the valence electrons of the gas medium. As a consequence, most of the laser emission wavelengths in this type of laser are produced in the visible part of the spectrum.

One of the most common and familiar examples of the neutral atom gas laser is the He–Ne laser. Examples of ionic gas lasers are the Ar^+ or Kr^+ lasers. The particular oscillating transitions and operation mechanisms can be found elsewhere (Siegman, 1986).

Figure 2.10 shows the relative laser output power of Ar^+ and Kr^+ lasers. As shown, gas lasers can oscillate simultaneously on various transitions. Single-line oscillation is usually carried out by using cavity mirrors with a maximum reflectivity at the desired wavelength. In the case of broadband reflectors, although continuous tuning is not possible, a certain wavelength selection is achievable among the lines of the discrete emission spectrum of the atoms in the gas.

In gases (atomic or ionic) the electronic energy levels of free atoms are narrow, since they are diluted systems and perturbation by the surroundings is very weak. An important fact derived from the discrete nature of the electronic levels in a gas is the high monochromaticity of the laser lines in this type of laser, compared to that of solid-medium based lasers. The high degree of coherence achievable with gas lasers is also a characteristic feature related to the narrow linewidth.

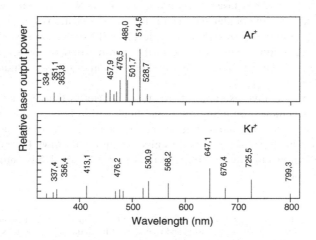

Figure 2.10   The spectral dependence of the laser output power of Ar^+ and Kr^+ lasers.

Usually, mainly Doppler broadening determines the gain profile of a particular laser transition. Indeed, due to the different configurations achievable with gas lasers (namely, a large cavity length), the laser line can be narrower than the Doppler linewidth. Different experimental realizations of single-mode lasers are detailed elsewhere (Demtröder, 2003).

EXAMPLE 2.4   *Obtain the separation between adjacent modes in a He–Ne laser whose cavity length is 1 m. The emission wavelength is* $\lambda = 632.8$ *nm.*

Different axial modes (optical standing waves) can be set up in a cavity of length L provided that the frequency ν fulfills the condition

$$\nu = \frac{nc}{2L}$$

n being an integer. Hence, the separation between adjacent modes is given by

$$\Delta\nu = \frac{c}{2L}$$

For $L = 1$ m, the separation between adjacent modes is $\Delta\nu = 150$ MHz.

The Doppler broadening of the Ne $3s_2 \rightarrow 2p_4$ transition, which is responsible for the laser emission at 632.8 nm at room temperature (by far the most important broadening mechanism) is given by $\delta\nu_d \sim 1500$ MHz.[2] Thus, the width of the laser gain profile may be of the same order of magnitude, ~ 1000 MHz. Comparing this data with the result obtained in our example, we can assert that several axial modes will attempt to resonate in the laser cavity; specifically, seven in our case.

(b) The other type of gas laser refers to the version in which the active medium is constituted by a molecular gas. Here, the laser transitions take place between rotational–vibrational *(ro-vibrational)* levels of the molecular system. Therefore, the laser emission wavelengths lie in the infrared part of the spectrum, corresponding to processes of activation or deactivation of rotations and/or vibrations in a molecule.

The representative example of this type of laser is the CO_2 laser. Laser oscillation is achieved on many rotational lines within two vibrational transitions of the molecule. Without any line selection, the system oscillates at only around 10.6 μm. This transition is found to give a continuous wave (cw) output power of several kilowatts with an efficiency of around 30 %, which is quite exceptional for a gas laser.

---

[2] The Doppler width is given by $\delta\nu = 2\nu_0 \left[(2\kappa T \ln 2)/Mc^2 \right]^{1/2}$ where k is the Boltzmann constant, T is the temperature, M is the mass of the atom, and ν_0 is the central frequency of the transition.

Due to the narrow absorption lines of gases, optical pumping is not efficient and electrical pumping is needed. Thus, discharge tubes containing the active gas are used. In their principal aspects, discharge gas laser tubes are similar for all gas lasers. Usually, they are provided with Brewster-angle end windows, which transmit light of the proper polarization. In the case of neutral atom gas lasers, such as the He–Ne laser, typical DC power inputs are of the order of 10 W. However, the pump process in an ion gas laser requires a high pumping power (several kilowatts), since it is a two-step process: in the first step, ions are produced from neutral atoms by collisions with electrons from the discharge; in the second, the ions thus formed are excited to a higher energy level. Due to the use of a high-density current, efficient cooling systems are required.

Gas lasers operate mainly in the cw regime, although in the case of singly ionized gases, both the pulsed and the continuous regime are used.

The characteristics of ionic or neutral atom gas lasers (namely, directionality, narrow linewidth, and high coherence length) have determined the great variety of scientific applications in which they have been used up to now. They have served as wavelength and frequency standards, and as alignment systems, and they have been an important tool in holographic experiments.

2.4.3 Dye Lasers

In this type of laser, the active media are organic dye molecules dissolved in a liquid (such as ethyl or methyl alcohol), which display strong broadband fluorescence spectra under excitation by visible or UV light.

A simplified energy-level diagram for a dye in a liquid solution is shown in Figure 2.11(a). The dye molecules have singlet (as well as triplet) electronic states.[3] Each electronic state comprises several vibrational states, and each of these in turn contains several rotational levels. When the dye molecules are excited with photons of the appropriate energy, higher vibrational levels of the first or second excited singlet states (S_1 or S_2) are populated from rotational–vibrational electronic (ro-vibronic) levels of the ground state S_0. Then, induced by collisions with the solvent, the excited molecules undergo a rapid nonradiative relaxation to the lower vibronic level of the S_1 state. At a sufficiently high pump intensity, population inversion may be achieved between the level v_0 in S_1 and the higher ro-vibronic levels v_k of S_0. Therefore, gain occurs on the transition $v_0(S_1) \rightarrow v_k(S_0)$.[4] The pumping cycle can be described by a four-level system, since the lower laser level $v_k(S_0)$ is depleted by collisions with the solvent molecules. Figure 2.11(b) shows the absorption and emission spectra of

---

[3] Singlet and triplet states refer to those states that have a total electron spin quantum number equal to 0 and 1, respectively. The spin multiplicity calculated as $2S + 1$, where S corresponds to the spin angular momentum, corresponds to 1 and 3 for the singlet and triplet states, respectively.

[4] A further possibility is that molecules undergo a nonradiative transition from the excited singlet states to the triplet states (intersystem crossing). This is the origin of some loss mechanisms for dye lasers.

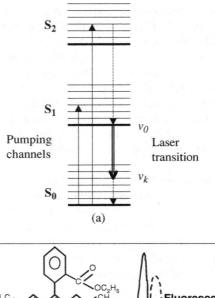

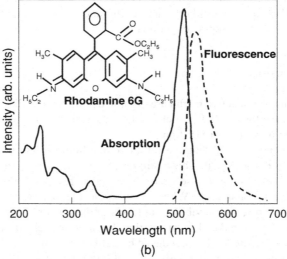

Figure 2.11 (a) An energy-level scheme in dye molecules. (b) The absorption and fluorescence spectra of Rhodamine 6G, together with its molecular scheme (reproduced with permission from Demtröder, 2003).

a commonly used dye, Rhodamine 6G. A representation of this dye molecule is also shown. As for all dye molecules, it has hexagonal carbon ring structures. These rings have associated loosely bound electrons that can move between different nuclei within the plane of the ring. These electrons provide the energy-level structure of the dye molecules.

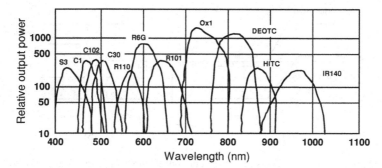

Figure 2.12 Tuning ranges for different dye lasers, illustrating the spectral regions covered by these lasers (pulsed condition) (courtesy of Spectra Physics).

The essential characteristic of dye lasers is their broad homogeneous gain profile. This feature has two important consequences:

• On the one hand, the output wavelength of a dye laser can be continuously varied within their broad emission band (various tens of nanometers). Therefore, with different dyes the overall spectral range covered by these lasers can be extended from around 400 nm to 1.1 μm, as shown in Figure 2.12.

• On the other hand, the homogenous broadening allows all of the excited molecules to contribute to the gain at a given frequency. This implies that under single-mode operation the output power should be not much lower than the multimode power, provided that the selecting intracavity elements do not introduce large additional losses.

Various designs for dye lasers, with different pumping sources and geometries, resonator configurations, and flow systems, have been successfully used to optimize laser performance.

Concerning pumping sources, the experimental realizations of dye lasers use flash-lamps, pulsed lasers, or cw lasers in various geometries. Since the absorption bands of many dyes reach the near UV region, the N_2 laser at 337 nm is well suited as a pumping source. Sometimes frequency-doubled pulsed neodymium YAG or cw Kr^+ and Ar^+ lasers are used, taking advantage of the absorption bands in the blue–green spectral region.

Dye lasers have been one of the most widely used types of tunable laser. In pulsed conditions, typical peak powers are in the range of 10^4–10^6 W. In the cw regime, reported powers are in the order of watts, with linewidths of around 1 MHz. Due to their flexibility in design and performance, dye lasers have been commonly used in a great variety of spectroscopic techniques, including high-resolution spectroscopy.

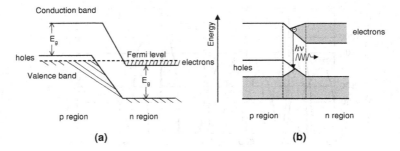

Figure 2.13   The energy levels of a p–n junction: (a) unbiased; (b) with a forward voltage applied.

2.4.4 Semiconductor Lasers

Semiconductor lasers have undergone a considerable metamorphosis during the past 30 years. They have grown and developed into a whole range of sophisticated opto-electronic devices. It is beyond the scope of this section to give a detailed description of the different semiconductor lasers, but we shall summarize the basic principles of this type of laser.

Figures 2.13(a) and 2.13(b) illustrate the basis of a semiconductor diode laser. The laser action is produced by electronic transitions between the conduction and the valence bands at the p–n junction of a diode. When an electric current is sent in the forward direction through a p–n semiconductor diode, the electrons and holes can recombine within the p–n junction and may emit the recombination energy as electromagnetic radiation. Above a certain threshold current, the radiation field in the junction becomes sufficiently intense to make the stimulated emission rate exceed the spontaneous processes.

The radiation can be amplified by an optical resonator, which, in the simplest case, is constituted by the semiconductor itself, shaped in the appropriate manner; for instance, by cutting the crystal so that two end faces are parallel to each other, and exactly perpendicular to the laser beam emitted by the junction (see Figure 2.14).

In semiconductor lasers, the emission wavelength is essentially determined by the band gap of the material, which governs the gain profile. Nowadays, by choosing the proper semiconductor system, a very wide spectral region can be covered by these lasers (0.37–5 μm), as can be observed in Figure 2.15.

An important characteristic of this type of laser is the possibility of some degree of continuous tunability over part of the gain profile. For wavelength tuning, all of those parameters that determine the energy gap between the upper and lower laser levels may be varied. In particular, a temperature change produced by an external cooling system or by a current change is frequently used to generate a wavelength shift. Typical tuning in the near infrared and visible range is of the order of several nanometers.

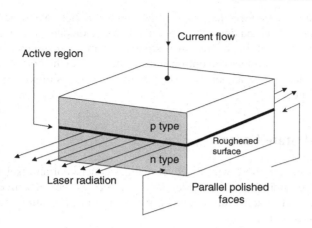

Figure 2.14 The structure of an injection laser.

Semiconductor lasers can operate in the cw regime (with output powers ranging from μwatts to tens of watts) or in the pulsed regime, with typical peak powers of tens of watts.

Semiconductor-based lasers have been further developed from the simple model depicted in Figure 2.14. The predictions by Kroemer and Alferov in the early 1960s stated that the concentrations of electrons, holes, and photons would become much higher if they were confined to a thin semiconductor layer between two other layers (Kroemer, 1963). Since then, sophisticated configurations for semiconductor heterostructures lasers have been made possible due to the development of fabrication techniques (Wilson and Hawkes, 1998; Kasap, 2001).

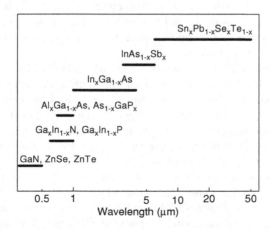

Figure 2.15 The spectral range of laser emission for several semiconductor materials.

We can already deduce that, due to the characteristics of the active medium, compact and miniaturized devices are attainable for semiconductor lasers. This fact, together with the possibility of custom-designed systems, constitutes a real advantage from the viewpoint of integrated opto-electronic devices. In the field of spectroscopy, they are commonly used as pumping sources for other types of solid state lasers, as will be seen later.

2.4.5 Solid State Lasers

Solid state lasers are those whose active medium consists of an insulating material activated by an optically active center. Three different types of active center have usually been used as active laser centers: rare earth ions, transition metal ions, and color centers (see Chapter 6).

When designing a new solid state laser system, an appropriate choice of the matrix – active center combination is needed. On the one hand, the active center should display optical transitions in the transparency region of the solid, which consequently requires the use of wide-gap materials. Additionally, the transitions involved in the laser action should show large cross sections in order to produce efficient laser systems. This aspect, which is directly related to the transition probability, is treated in depth in Chapters 5 and 6, where the physical basis of the behavior of an optically active center in a solid is studied.

There are many solid state lasers. One of the most commonly treated types in laser textbooks is the ruby laser (Al_2O_3:Cr^{3+}), which was the first laser system demonstrated by T. H. Maiman at the Hughes Research Laboratory early in 1960 (Maiman, 1960). Figure 6.9 in Chapter 6 will show the quantum energy levels associated with the unfilled 3d inner shell of the Cr^{3+} ion when it substitutes for the Al^{3+} ion in the Al_2O_3 lattice crystal. By using a ruby rod placed inside a spiral flashlamp filled with a hundreds of torrs of xenon, it is possible to optically pump Cr^{3+} ions from the $^4A_{2g}$ ground state into the broad $^4T_2$ and $^4T_1$ bands of the excited levels. After a rapid relaxation down to the very sharp ^2E_g level, laser emission can be produced at 694 nm via the $^2E_g \rightarrow {}^4A_{2g}$ transition.

This laser is then a three-level laser, in which the lower laser level is also the ground energy level. This is usually very unfavorable, but with sufficient pumping power it is possible to produce enough population inversion and a high laser output power, in the range of 50 mJ to 0.5 J per shot (Q-switch regime). This laser has shown a wide utility in a variety of process, including rapid thermal annealing of materials, but nowadays its use is quite limited, and it has been practically substituted by new laser devices that offer better efficiencies and performance.

Some of the most commonly used solid state lasers by far are the Nd^{3+} based lasers, such as the Nd:YAG laser (Nd^{3+} ions in yttrium aluminum garnet), Nd:glass materials, or more recently the Nd:YLF or Nd:YVO$_4$ lasers. Nd^{3+} lasers operate in a four-level scheme and they are optically pumped either by a flashlamp or, for a more

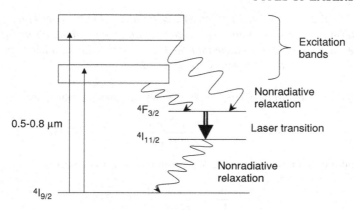

0.5-0.8 μm

$^4F_{3/2}$

$^4I_{11/2}$

$^4I_{9/2}$

Excitation bands

Nonradiative relaxation

Laser transition

Nonradiative relaxation

Figure 2.16   The Energy levels and operational scheme of Nd^{3+} based solid state lasers.

compact and efficient system, by semiconductor lasers. The energy-level diagram and operational scheme is depicted in Figure 2.16.

The upward arrows in the figure indicate the pumping channels to various high energy levels by flashlamp (0.5 μm) or semiconductor lasers (0.8 μm), where Nd^{3+} ions display strong absorption transitions. The downward arrow indicates the widely used laser emission at 1.06 μm, associated with the $^4F_{3/2} \rightarrow {}^4I_{11/2}$ transition. In addition, laser action is also generally possible from the same $^4F_{3/2}$ level to the $^4I_{9/2}$ state at around 0.9 μm and to the $^4I_{13/2}$ state at around 1.3 μm.

Compact and stable devices are available that take advantage of the improved quality of the crystal lasers, as well as increased pump efficiencies. Hundreds of different models of Nd^{3+} based lasers have demonstrated laser action (Kaminskii, 1981). It is possible to operate these Nd^{3+} solid state lasers in the continuous regime, with output powers ranging from 1 W to 1000 W. Pulsed operation is also possible, with a pulse length from the picosecond range, via mode-locking, to tens of nanoseconds by Q-switch operation.

Nd lasers, and more specifically the Nd:YAG laser, cover a very broad range of applications, from a number of small-scale laser cutting, drilling, and marking applications to an enormous variety of scientific and technological experiments. The high output powers delivered by these lasers make them very useful tools to generate the intense radiation needed for a variety of nonlinear processes, such as harmonic generation, parametric processes, or stimulated Raman scattering. For instance, it is very common to find Nd^{3+} laser systems in which not only the laser line at 1.06 μm is supplied, but also coherent radiation at 532 nm, 355 nm, 266 nm, or even 213 nm, corresponding to the second, third, fourth, and fifth harmonic generation of the fundamental laser emission, respectively. This used to be performed by employing nonlinear crystals, such as KDP or BBO, placed inside or outside the laser cavity.

EXAMPLE 2.5   *For a Nd:YAG laser crystal located in a cavity of length 0.1 m, determine (a) the frequency separation among the axial modes and (b) the number of axial modes. Assume that the laser is operating in a laser line at 1.06 μm, whose full width, δv, is 1.1 × 10¹¹ Hz.*

(a) The frequency separation, Δv, between two adjacent modes is

$$\Delta v = \frac{c}{2L} = \frac{3 \times 10^8 \,\mathrm{m\,s}^{-1}}{2 \times 0.1\,\mathrm{m}} = 1.5 \times 10^9 \,\mathrm{Hz}$$

(b) The number of axial modes, N, within the laser linewidth is given by

$$N \approx \frac{\delta v}{\Delta v} = \frac{1.1 \times 10^{11}}{1.5 \times 10^9} \approx 73 \,\mathrm{modes}$$

2.5 THE TUNABILITY OF LASER RADIATION

Nowadays, tunable coherent radiation has become a fundamental tool in many areas (remote sensing, isotope separation, photochemistry, etc.), with particular relevance in optical spectroscopy.

Tunable coherent light sources can be realized in several ways. One possibility is to make use of lasers that offer a large spectral gain profile. In this case, wavelength-selecting elements inside the laser resonator restrict the laser oscillation to a narrow spectral interval and the laser wavelength may be continuously tuned across the gain profile. Examples of this type of tunable laser are the dye lasers were treated in the previous section.

Another possibility for laser wavelength tuning is based on the shift of the energy levels in the active medium by external perturbations, which can cause a corresponding spectral shift in the gain profile and, therefore, in the laser wavelength. For instance, this shift may be caused by a temperature variation, as in the aforementioned case of semiconductor lasers.

A third way of generating continuously tunable coherent radiation uses more complicated systems based on the principle of *optical parametric oscillation* (and amplification). Since the gain in these systems is not originated by stimulated emission, but by means of a nonlinear frequency conversion process, we will treat them in a separate section.

2.5.1 Tunable Solid State Lasers

We have previously mentioned some laser systems with which a certain degree of tunability is possible (dyes and semiconductor lasers). However, during the past two decades a new class of tunable lasers has been developed: *tunable solid state lasers*. They have definitively replaced dye lasers for some spectral regions, offering a higher degree of compactness, a higher output power with lower pumping powers, and the absence of bleaching or aging effects in normal operation. Compared to semiconductor lasers, tunable solid state lasers, can offer a much larger tunable range with a higher frequency stability and a better quality beam.

In general, two types of tunable solid state lasers have been developed: those based on *color centers* in alkali halide crystals, and those based on *transition metal ions* (3d) in a crystalline host. In both cases, the tunability relies on the large spectral gain profile provided by the active center.

Color centers in alkali halide crystals are based on halide ion vacancies in the crystal lattice, as will be explained in detail in Chapter 6. When a single electron is trapped at such a vacancy, its energy levels result in new absorption and emission lines, broadened into bands by interactions with phonons. The phonon-broadened emission bands are the origin of the tunability of these lasers. Although an overall spectral range between 0.85 and 3.6 μm could be covered by diverse color centers in different alkali halide hosts, these lasers show a main drawback: they need to operate at low temperatures in order to achieve a high quantum efficiency. The crystal should be mounted on a cold finger cooled with liquid nitrogen, which leads to complex cavity designs. This has strongly limited their use and, at the present time, they have been almost completely substituted by other systems with more practical designs (mainly by those involving frequency-mixing techniques).

Let us now confine our attention to *transition metal ion-activated tunable solid state lasers*. Similar to the case of color center lasers, tunability in 3d ion-activated lasers is due to the fact that the energy involved in the laser transition is distributed into the creation of photons and phonons. That is, the lower laser level is not a well-defined electronic level but a 'band' of vibrational states and, thus, in the laser transitions not only there is a change in the electronic energy of the active center, but also a change in the energy of the vibrational modes of the crystalline host. This explains why these lasers are called 'vibronic lasers' or 'phonon-terminated lasers.' Figure 2.17 shows a schematic energy-level diagram for this kind of laser. As observed, these devices act as four-level lasers.

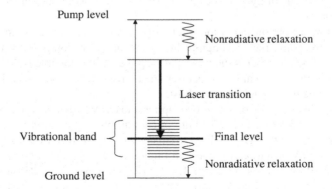

Figure 2.17   A simplified energy-level diagram for a vibronic laser.

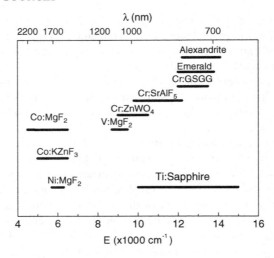

Figure 2.18   The spectral range covered by different tunable solid state lasers based on transition metal ions.

Chapter 5 will show in more detail how the spectral width of optical transitions of active centers (particularly for transition metal ions) is affected by lattice vibrations. For the purpose of this section, we will just mention that these transitions are associated with the outer electrons of the active center (the 3d valence electrons), which show strong interactions with the phonons of the matrix in which they are embedded. As a result, the optical transitions, and particularly the emission lines, are strongly modulated by lattice vibrations.

Figure 2.18 shows the range covered by different tunable solid state laser systems based on transition metal ions. As observed, a good variety of matrices have shown tunable laser action on the basis of Cr^{3+} as an active ion. The fundamental aspects determining the tunability of those Cr^{3+} based systems will be the subject of Section 6.4 in Chapter 6.

Let us now devote our attention to one of the most popular lasers in the field of optical spectroscopy: the Ti–sapphire laser. As shown in Figure 2.18, the spectral range covered by this laser is the largest among the various tunable solid state lasers: from 675 nm up to 1100 nm. In this laser, the active medium is formed by optically active Ti^{3+} ions in the Al_2O_3 crystal host.

Ti^{3+} ions in Al_2O_3 shows a broad optical absorption band at around 500 nm (see Figure 6.7). This ensures efficient optical pumping by a variety of sources, the most common being the Ar^+ gas laser, frequency-doubled Nd-based solid state lasers (Nd:YAG, Nd:YVO$_4$, etc.), or green-emitting diode lasers.

Figure 2.19 shows, as an example, the output power of a Ti–sapphire laser pumped by a 15 W multiline Ar^+ laser. The wide spectral region covered by this laser substitutes for those of the dye lasers emitting in the same spectral region. Several sets

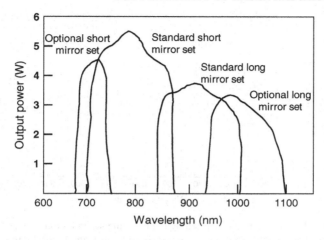

Figure 2.19   The spectral output power profile of an Ar^+ pumped Ti–sapphire laser (courtesy of Spectra Physics).

of broadband reflectors (including input and output mirrors) should be used to cover the whole spectral gain region. For wavelength selection, intracavity elements, such as a birefringent filter (a Lyot filter), are needed (Kobtsev and Sventsitskay, 1992).

The Ti–sapphire laser is nowadays one of the fundamental tools in a great variety of studies concerning solid state spectroscopy. In fact, it constitutes one of the most versatile excitation sources, showing valuable characteristics in addition to that of a large tuning range. For example, it can work in the cw regime in single-mode operation with an extremely good beam quality, while in the pulsed regime it can operate with pulse lengths as short as 12 fs. The possibility of compact and stable designs is also a very important characteristic of Ti–sapphire laser devices. As will be shown in the next section, these lasers are usually the basis of more complicated systems based on frequency-mixing techniques, which take advantage of the good characteristics provided by the radiation of the Ti–sapphire laser.

2.5.2 Tunable Coherent Radiation by Frequency-Mixing Techniques

We have previously mentioned how from Nd^{3+} based lasers emitting in the near infrared region, two, three, or even four wavelengths in the visible and UV regions can be coherently generated by using the appropriate nonlinear crystals. This is an example of the potentiality of the frequency-mixing techniques that can be used, taking advantage of the intense radiation provided by laser devices.

In a general way, when the radiation from two laser beams with frequencies ω_1 and ω_2 are superimposed in a medium with a sufficiently large nonlinear part of the susceptibility, each atom is induced to undergo forced oscillations and generates

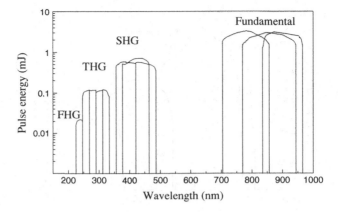

Figure 2.20   The spectral range covered by the fundamental radiation of a Ti–sapphire laser and the various harmonic generation processes: second, third, and fourth harmonic generation (SHG, THG, and FHG, respectively) (courtesy of Quantronix).

radiation with the sum frequency $\omega_1 + \omega_2$ and the difference frequency $\omega_1 - \omega_2$. In other words, nonlinear optics, in the form of *three-wave interactions*, supplies us with a means to merge and split photons. Basically, there are only two three-wave interaction processes: *fusion* $\omega_1 + \omega_2 \rightarrow \omega_3$ and *fission* $\omega_3 \rightarrow \omega_1 + \omega_2$ of photons. Of course, the resultant wave is coherently constructed for a particular direction in the nonlinear medium for which the phase velocities of primary and secondary waves are identical. In that case, we say that the *phase-matching* condition is fulfilled. If the frequencies ω_1 or ω_2 of the incident radiation can be tuned, the difference or sum frequency exhibits the same absolute tuning range, provided that the phase-matching condition is always fulfilled over that tuning range. As an example, Figure 2.20 shows the broad spectral ranges that can be covered by using Ti–sapphire as a fundamental radiation for frequency conversion to the second, third, and fourth harmonic outputs (205–1000 nm).

Therefore, it is clear that different frequency conversion processes, together with the variety of lasers, have led to a great variety of coherent sources with large and differing tuning ranges. The whole spectral range of 185–3400 nm can be covered by different methods.

We will restrict our description to those processes and systems that have been developed more over recent years: those based on the *optical parametric oscillation* or *amplification* processes. The practical devices involve two of the most relevant solid state lasers nowadays: the Nd:YAG laser and the Ti–sapphire laser.

2.5.3 Optical Parametric Oscillation and Amplification

The *optical parametric oscillator* (OPO) is based on the parametric interaction of a strong pump wave with a nonlinear medium that has a highly nonlinear susceptibility.

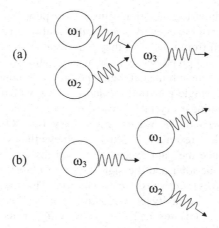

Figure 2.21   Different types of three-wave interaction in a nonlinear media (at the photon level): (a) sum-frequency generation and (b) the optical parametric generation.

The interaction can be described as an inelastic scattering of a pump photon, $\hbar\omega_p$, by the nonlinear medium, in which the pump photon is absorbed and two new photons, $\hbar\omega_s$ and $\hbar\omega_i$, are generated. The subscripts s and i stand for *signal* and *idler* (photons or waves).

Figure 2.21 shows an illustration, at the photon level, of the parametric processes compared to the sum-frequency generation process.

In the parametric oscillation, energy conservation requires that

$$\omega_p = \omega_i + \omega_s \tag{2.9}$$

where ω_p, ω_i and ω_s are represented by ω_3, ω_1 and ω_2, respectively, on the Figure. Similarly to the sum-frequency process, the parametrically generated photons ω_i and ω_s are coherently added if the phase-matching condition

$$\vec{\kappa}_p = \vec{\kappa}_i + \vec{\kappa}_s \tag{2.10}$$

is fulfilled, where $\vec{\kappa}_p$, $\vec{\kappa}_s$, and $\vec{\kappa}_i$ are the wavevectors of the pump, signal, and idler, respectively. Equation (2.10) corresponds to the conservation of momentum for the photons involved in the process. Both conditions (2.9) and (2.10) can be satisfied with suitable nonlinear media. The most efficient generation is achieved for collinear phase-matching, where $\vec{\kappa}_p \parallel \vec{\kappa}_s \parallel \vec{\kappa}_i$. For this case, we have

$$n_p\omega_p = n_s\omega_s + n_i\omega_i \tag{2.11}$$

where n_p, n_s, and n_i correspond to the refractive indexes for the pump, signal, and idler interacting waves.

Simply stated, parametric generation splits a pump photon into two photons, which satisfies energy conservation at every point in the nonlinear crystal. The frequencies ω_i and ω_s are initiated from noise and they only have to obey the conditions (2.10) and (2.11). That is, for a given wavevector of the pump wave, the phase-matching condition selects, out of the infinite number of possible combinations $\omega_i + \omega_s$ allowed by Equation (2.10), a single pair $(\omega_i, \vec{\kappa}_i)$ and $(\omega_s, \vec{\kappa}_s)$, which is determined by the orientation of the nonlinear crystal with respect to $\vec{\kappa}_p$. The resultant macroscopic waves are called *signal* and *idler* waves, respectively. In OPO systems, the nonlinear crystal is placed inside a resonator in order to achieve the oscillation of the idler or signal frequencies, once the gain exceeds the total losses. The optical cavity may be resonant for both the idler and the signal waves (*doubly resonant oscillator*) or only for one of the waves (*singly resonant oscillator*). The first type requires a single-frequency pump source and cavity length control to ensure that both parametric waves are at resonance. The singly resonant cavity has a higher threshold, but it does not require length control and it is simpler to design.

A different approach, that lacks the presence of a resonant cavity, is constituted by the so-called *optical parametric amplifier* (OPA). In absence of a resonator, the three wave interaction processes in a nonlinear medium may be used to amplify a light signal at a frequency ω_s. To this end, an intense pump wave of frequency ω_p, such as $\omega_p > \omega_s$, must be applied to the nonlinear crystal together with the signal. Then, provided that conditions (2.9) and (2.10) are satisfied, an additional wave is generated at $\omega_i = \omega_p - \omega_s$, the idler wave. The signal wave is amplified, whereas the idler wave is newly generated. Figure 2.22 shows an illustration of this process compared to the OPO. The OPA is a real signal amplifier that can be used for generating coherent light at new frequencies that are not bound to any atomic transition. In fact, this is the basis of the operation of the OPO, since an amplifier for a certain frequency can always also be used for the construction of an oscillator. In this case, all that has to be done is to provide phase-matched feedback and to overcome the unavoidable losses.

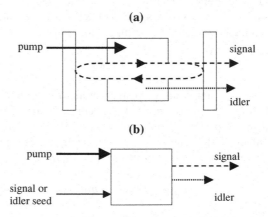

Figure 2.22 Schematic representations of (a) OPO and (b) OPA processes in a nonlinear crystal.

EXAMPLE 2.6 *In an OPO system, the pump beam wavelength is set at 355 nm. Assuming that the signal beam is produced at 450 nm, determine the wavelength corresponding to the idler beam.*

The wavenumbers of the pump, $\tilde{\nu}_p$, and signal, $\tilde{\nu}_s$, beams are obtained from the aforementioned wavelengths:

$$\tilde{\nu}_p = 28\,169\,\text{cm}^{-1} \quad \text{and} \quad \tilde{\nu}_s = 22\,222\,\text{cm}^{-1}$$

Therefore, the wavenumber of the idler beam is

$$\tilde{\nu}_i = \tilde{\nu}_p - \tilde{\nu}_s = 5947\,\text{cm}^{-1}$$

which corresponds to a wavelength of

$$\lambda_i = \frac{1}{5947 \times 10^{-7}\text{nm}^{-1}} = 1681\,\text{nm}$$

With optical parametric devices, coherent light production can be tuned in frequency (or wavelength) by changing the feedback conditions. On the other hand, any frequency from a continuous range of frequencies may be generated, depending on the geometric setting. Additionally, for a fixed pump frequency ω_p, tuning of the OPO and of OPA systems can be accomplished by any process that changes the refractive indexes at the pump (ω_p), signal (ω_s), or idler (ω_i) frequency. These processes include crystal rotation or controling the crystal temperature. Among the most common nonlinear crystals used in optical parametric devices are beta barium borate (BBO) crystals, potassium dihydrogen phosphate (KDP) crystals, and lithium niobate crystals. They show highly nonlinear coefficients and excellent tuning ranges. Continuous coverage from the far infrared to the ultraviolet region is possible in OPO- or OPA-based devices. It should be mentioned that these OPOs and OPAs are nowadays in most cases all solid state devices, which implies stability and compactness.

Figure 2.23 shows the broad spectral region covered by an OPO system, using as a pump wavelength the Q-switched radiation at 355 nm from a system based on a Nd:YAG laser. Coherent radiation that is tunable from 400 nm to 2 μm can be obtained from the signal and idler waves.

2.6 ADVANCED TOPICS: SITE SELECTIVE SPECTROSCOPY AND EXCITED STATE ABSORPTION

We will briefly describe the basis of two spectroscopic methods that are commonly used in the investigation of optically active centers in solids. As we will see, both of them constitute good examples of the usefulness of laser properties in the field of spectroscopy.

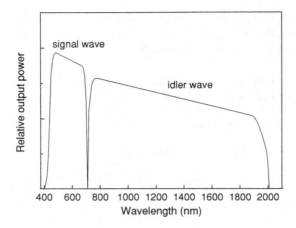

Figure 2.23   The spectral region covered by an OPO system, using as a pumping source the radiation at 355 nm from a Nd:YAG laser (courtesy of Spectra Physics).

2.6.1 Site Selective Spectroscopy

As was mentioned in Chapter 1 (Section 1.3), the optical absorption and emission lines of active ions in solids show, as an important broadening mechanism, so-called inhomogeneous broadening (or strain broadening). The presence of a distribution of slightly different crystal field environments – due, for instance, to imperfections in the crystal lattice (growth strains, dislocations, faults, unintentional impurities, etc.) – alters the energy-level structure of the optically active ions. In the simplest case, the inhomogeneous linewidth consists of a broad envelope, given by Equation (1.9), which represents a probability distribution of homogeneous lines spread in energy by those variations in the crystal field at the different optically active impurity sites. This inhomogeneous broadening can range from a small fraction of a wavenumber for the best crystals to many hundreds of wavenumbers for amorphous hosts.

However, the inhomogeneous effects are not limited to the linewidth of the transitions. In many materials, for example, the active ions can substitute into different and distinct crystallographic lattice sites, each of which can be described by its own particular parameters and symmetry. The resulting spectrum will be a composite of the spectra of the active ion at these sites. As a result, it can be highly complex due to the contribution of the different *nonequivalent optical centers* in the host.[5] The nonequivalent environments may arise from different sources, among which we should mention disorder in the host crystal, different local charge compensation mechanisms, concentration effects, or codoping (García Solé and Bausá, 1995).

In some cases, the center structure can be directly observed in the low-temperature absorption/emission spectra, but laser radiation is needed to unravel the spectral

---

[5] The term 'center' should be understood here as an optically active ion in a specific crystal environment.

features of a particular center. Wright and co-workers were the first to use high-resolution selective laser excitation to solve such complex spectra and to determine the symmetries and composition of the individual centers. Their initial work was performed on Er^{3+}:CaF_2 (Tallant and Wright, 1975). They used a high-resolution tunable dye laser as excitation source to *selectively populate* only one type of Er^{3+} center at a time. The simplified fluorescence from each center was then measured in turn. From the energy-level structure of each individual center, information about its nature can be derived.

This technique constitutes a good example of high-resolution laser spectroscopy. It has been successfully applied to a variety of systems to examine important aspects, such as the microscopic crystalline structure, the trace impurity distribution, or the degree of structural disorder.

2.6.2 Excited State Absorption

As was shown in Chapter 1 (Section 1.3), the absorption spectrum is a useful tool that provides information on the transitions taking place from the ground level to the different excited levels of a given system. However, in certain practical cases it could also be interesting to acquire information on the nature and intensity of the absorption transitions that could take place among different excited states. In this sense, the aim of excited state absorption spectroscopy (ESA) is to quantify the absorption from an excited state of a system up to its higher excited levels. Under equilibrium conditions, the absence of a population in the departing excited state does not permit the absorption process and accordingly, the experimental setup used to characterize and quantify the ESA process needs two light sources. On the one hand, a laser beam (the *pump* beam) is used to achieve a large population density in the selectively excited state from which ESA takes place. On the other hand, and simultaneously, a broadband source, either a tunable laser or a lamp, is used as a *probe* to perform a wavelength scan to obtain the optical absorption from the highly populated excited level (the initial level) to the higher excited levels (the terminal levels). The experiments can be carried out by measuring the intensities transmitted by the sample with and without the pump beam, as a function of the scan wavelength of the probe beam. The difference in the absorption coefficients of the sample with and without the pump beam ($\alpha' - \alpha$) is then directly obtained as follows:

$$\alpha' - \alpha = \ln(I_u/I_p)d \qquad (2.12)$$

where I_u is the transmitted intensity without the pump beam, I_p is the transmitted intensity with the pump beam, and d is the sample thickness. Provided that the absorption coefficient α is known, from ESA measurements it is possible to obtain the absorption coefficient for the excited state absorption process.

More sensitive than ESA is so-called excited state excitation (ESE), by using fluorescence spectroscopy. In this case, the excitation spectrum corresponding to an emission from a high-lying level is monitored as a function of the frequency of the tunable laser probe. The obtained ESE spectrum has the same profile as the ESA spectrum but needs to be properly calibrated in terms of cross section units (Guyot and Moncorge, 1993).

The experiments have demonstrated the importance of characterizing ESA in order to assess the ability of a system as a laser material. For example, if the ESA process takes place in the spectral region of the laser emission, the laser gain can be strongly affected, and laser action could even be prevented. On the other hand, if ESA occurs in the pumping region of the laser material, the pump efficiency could be reduced, and ESA could also be a source of thermal loading for the laser system.

An additional advantage of ESA and ESE is the possibility of localizing high energy levels, which could not be accessed using conventional spectrophotometers. This is the case for some bands in the ultraviolet spectral region. Moreover, ESA could allow the observation of optical transitions, which are forbidden by one-photon spectroscopy (Malinowski *et al.*, 1994)

EXERCISES

2.1. The width of the gain profile in a CO_2 laser is given as 66 MHz (close to the Doppler width of the emission band of the gas). If the eigenfrequency of the laser resonator is tuned to the center of the laser gain profile, what is the maximum length of resonator for which the laser can oscillate in a single mode?

2.2. Considering that only reflection losses due to the mirrors of the cavity cause the decrease of the energy stored in the resonator modes, determine the expression for the mean lifetime of a photon in the resonator as a function of the reflectivities of the mirrors, R_1 and R_2.

   Note that the mean lifetime of a photon in the cavity can be regarded as the time during which the energy stored in the corresponding mode has decreased to $1/e$ of its value at $t = 0$.

2.3. The output power of a cw Rhodamine 6G dye laser is mixed with selected lines of an Ar^+ laser, which is used to pump the dye laser with 15 W on all lines (see Figures 2.10 and 2.12). The superimposed beams are focused into a temperature-stabilized KDP crystal. Tuning is accomplished by simultaneously tuning the dye laser wavelength and the orientation of the KDP crystal. Determine the entire wavelength range that can be covered by using different Ar^+ lines.

REFERENCES AND FURTHER READING

Boyd, R. W., *Nonlinear Optics*, Academic Press, Inc., London (1992).
Demtröder, W., *Laser Spectroscopy*, 3rd edn, Springer Series in Chemical Physics 5, Springer-Verlag, Berlin (2003).

Duarte, F. J. and Hillman, L. H. (eds.), *Dye Laser Principles*, Academic Press, Inc., New York (1990).

García Solé, J. and Bausá, L. E., in *Insulating Materials for Optoelectronics*, ed. F. Agulló-López, World Scientific, Singapore (1995).

Guyot, Y. and Moncorge, R., *J. Appl. Phys.*, **73**, 8526 (1993).

Kaminskii, A. A., in *Laser Crystals*, ed. D. L. MacAdam, Springer-Verlag, Berlin (1981).

Kasap, S. O., *Optoelectronics and Photonics*, Prentice Hall, Upper Saddle River New Jersey (2001).

Kobtsev, S. M. and Sventsitskay, A., *Opt. Spectrosc.*, **73** (1) , 114 (1992).

Kroemer, H., *Proc. IEEE*, **51**, 1782, (1963).

Lauterborn, W. and Kurz, T., *Coherent Optics: Fundamentals and Applications*, 2nd edn, Springer Series in Advanced Texts in Physics, Springer-Verlag, Berlin (2003).

Maiman, T., *Nature*, **187**, 493 (1960).

Malinowski, M., Joubert, M. F., and Jaquier, B., *Phys. Rev. B*, **50**, 12 367 (1994).

Rodhes, C. K. (ed.), *Excimer Lasers*, Topics in Applied Physics, vol. 30, Springer-Verlag, Berlin (1979).

Shionoya, S. and Yen, W. M. *Phosphor Handbook*, CRC Press, Boca Raton, Florida (1999).

Siegman, A. E., *Lasers*, University Science Books, Mill Valley, California (1986).

Svelto, O. *Principles of Lasers*, 4th edn, Plenum Press, New York (1998).

Tallant, D. R. and Wright, J. C., *J. Chem. Phys.*, **63**, 2074 (1975).

Wilson, J. and Hawkes, J., *Optoelectronics: An Introduction*, 3rd edn, Prentice Hall, London (1998).

Yen, W. M. and Selzer, P. M. *Laser Spectroscopy of Solids*, 2nd edn, Topics in Applied Physics, vol. 49, Springer-Verlag, Berlin (1986).

3
Monochromators and Detectors

3.1 INTRODUCTION

A full understanding of the optical processes in solids requires the simultaneous combination of different experimental techniques devoted to the study of the intensity and spectral properties of the different light beams involved in optical spectroscopy experiments (i.e., excitation, reflected, transmitted, emitted, and scattered beams). In Chapter 2, the main light sources used in optical spectroscopic were introduced and their working principles were described. In this chapter, we will give a general overview of the different kinds of apparatus used to analyze the properties of the light beams interacting with the solid (intensity, spectral distribution, and temporal behavior). The different tools needed for this characterization are summarized in this chapter.

3.2 MONOCHROMATORS

The *monochromator* is a fundamental tool in optical spectroscopy. As we mentioned in Chapter 1, a monochromator is an optical element that is used to isolate the different spectral components of a light beam. Monochromators have two main utilities in optical spectroscopy experiments:

 (i) To transform the polychromatic beam generated by lamps into a monochromatic beam for selective excitation.

 (ii) To analyze the light emitted or scattered by any material after some kind of excitation (luminescence or Raman experiments). The emitted or scattered light

An Introduction to the Optical Spectroscopy of Inorganic Solids   J. García Solé, L. E. Bausá, and D. Jaque
© 2005 John Wiley & Sons, Ltd   ISBNs: 0-470-86885-6 (HB); 0-470-86886-4 (PB)

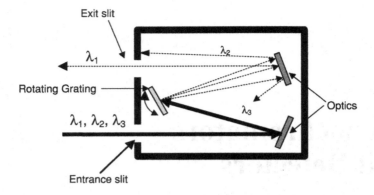

Figure 3.1 A schematic drawing of the simplest kind of monochromator.

usually extends over some range of the spectrum and the exact knowledge of its spectral distribution is required for a full understanding of the physical processes taking place after excitation.

Although in some cases monochromators are classified as *dispersive* or *nondispersive monochromators*, in this book we are only dealing with dispersive monochromators, since they are most commonly used in optical spectroscopy. In dispersive monochromators, a spatial separation is obtained for the different spectral components of the input beam. As shown in Figure 3.1, the simplest monochromators consist of the following elements:

(i) *A variable entrance slit.* The beam light to be analyzed is launched thorough it by using adequate optical elements.

(ii) *Monochromator optics.* These are used to image the entrance slit onto the exit slit. These optics usually consist of a set of mirrors.

(iii) *A dispersive element.* This could be a *prism* or a *diffraction grating*. In the first case, dispersion of the incoming light is produced by the wavelength dependence of the refractive index, whereas in the second case dispersion is produced as a consequence of interference effects. In general, grating monochromators show superior performance to prism monochromators, so in what follows we will only deal with grating monochromators.

(iv) *A variable exit slit.* The spectral component of interest (λ_1 in Figure 3.1) comes out of the monochromator thorough the exit slit. The spectral resolution of the monochromator, as well as the intensity of the outgoing light, depend on the width of the exit and entrance slits.

When a polychromatic beam reaches the grating, diffraction effects take place, so that the angle at which each spectral component is reflected depends on its particular

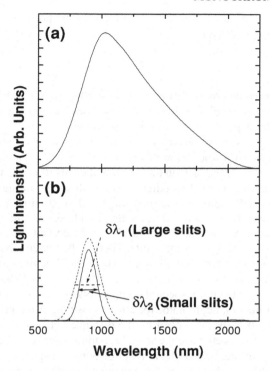

Figure 3.2 The spectral dependence of the light emitted from an incandescent lamp before and after it passes through the monochromator.

wavelength. The actual relationship between the reflected angle and the wavelength is given by the characteristics of the particular grating used (the *groove density*). As can be observed from Figure 3.1, for a fixed position of the grating only one spectral component (wavelength) of the incoming beam will reach the exit slit (λ_1 in our case). From Figure 3.1, it is also clear that the wavelength of the outgoing beam can be changed by simply rotating the grating.

As mentioned above, monochromators can be used to manipulate the spectral distribution of the light emitted by lamps. In order to illustrate this point, Figure 3.2 shows the spectral dependence of the light emitted by an incandescent lamp (Figure 3.2(a)) and the spectral dependence of the same beam after it has passed through a monochromator (Figure 3.2(b)). As can be observed, only one spectral component of the original beam is obtained at the exit slit of the monochromator. Of course, the output wavelength can be varied within the lamp emission range just by rotating the grating. We will now introduce the main parameters that are used to characterize any monochromator:

(i) *The spectral resolution.* This is the ability of the monochromator to separate two adjacent spectral lines. If the minimum spectral distance between two lines in the vicinity of λ that can be isolated by the monochromator is $\delta\lambda$, then the spectral resolution, R_0, is given by

$$R_0 = \frac{\lambda}{\delta\lambda} \qquad (3.1)$$

The resolution is determined by the spectral width of the output beam (monochromators with high resolution lead to narrow lines). In grating monochromators, the resolution depends on the number of grooves that exist in the grating (the resolution increases with the number of grooves), on the optical length traveled by the light beam inside the monochromator (longer monochromators lead to higher resolutions), and also on the width of the slits (the resolution increases as the slit width decreases). This latter aspect is illustrated in Figure 3.2(b), where the output signal is drawn assuming large and small slits. For large slits, the output spectrum becomes broader, leading to a larger $\delta\lambda$ and then reducing the monochromator resolution. It should also be noted that the output intensity is higher for larger monochromator slits. This can be understood by inspection of Figure 3.1; for a narrow exit slit, only one spectral component of the incident beam comes out of the monochromator (λ_1 in Figure 3.1). When the width of the exit slit is large, then more spectral components (λ_1 and λ_2 in Figure 3.1) could come out of the monochromator for a fixed position of the grating. Finally, the resolution is strongly influenced by the distance between the gratings and the slits; large distances enhance the spatial separation between the spectral components and, therefore, reduce the number of outgoing spectral components for a fixed slit width. For this reason, large monochromators are used in high spectral resolution experiments.

(ii) *Bandpass.* Monochromators are not perfect, as they produce an apparent spectral broadening for an ideal input beam that is purely monochromatic. Figure 3.3 shows the real spectrum (i.e., intensity versus wavelength) corresponding to a pure monochromatic beam (Figure 3.3(a)). If this beam passes thorough an ideal monochromator, its spectrum will not be changed (see Figure 3.3(b)). Nevertheless, real monochromators produce a different spectrum to that of monochromatic light, yielding a certain spectral (wavelength) dispersion (see Figure 3.3(c)). Thus the output line has a finite instrumental profile width. The full width at half maximum (FWHM) of the output beam is called the monochromator bandpass.

(iii) *The spectral response: 'blaze'.* As can be observed from a comparison of Figures 3.2(a) and 3.2(b), it is clear that the light intensity is reduced after it passes through the monochromator. Among the great variety of factors contributing to this reduction, the grating reflectance is the most important. The *blaze* wavelength is defined as the wavelength at which the grating operates with its highest efficiency. The efficiency of any holographic grating (the most common gratings used in conventional monochromators) is strongly dependent on the dispersed wavelength. Figure 3.4 shows the spectral response of two holographic gratings (in each case, the blaze wavelength is indicated by an arrow). As can be

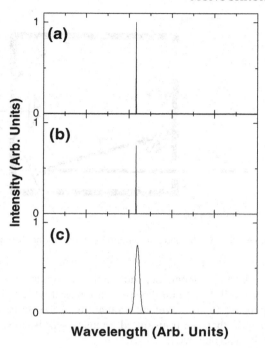

Figure 3.3 (a) The spectral distribution of a pure monochromatic beam. (b) The spectrum of this monochromatic beam after it passes through an ideal monochromator. (c) The spectrum of this monochromatic beam after it passes through a real monochromator.

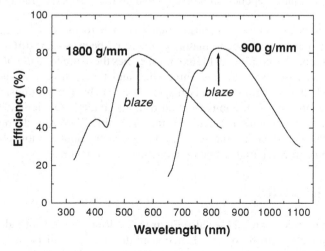

Figure 3.4 The spectral responses of two diffraction gratings with different groove densities.

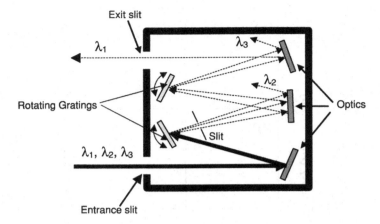

Figure 3.5   A schematic drawing of a double monochromator.

observed, the blaze wavelength depends on the groove density, so that the grating to be employed will depend on the spectral range at which the monochromator has to be operated. If the wavelength range of operation is far away from the blaze wavelength, then the transmitted intensity will be low, although spectral filtering will be also achieved.

(iv) *Dispersion*. This is associated with the ability of a given monochromator to produce, at the exit slit, a spatial separation between two adjacent wavelengths. It is given by $d\lambda/dx$, where dx is the spatial separation at the exit plane between two spectral lines whose wavelengths are λ and $\lambda + d\lambda$. The dispersion of a given monochromator depends on its length and on the particular grating used.

Finally, it is important to mention that we have focused our attention on the simplest version of a monochromator. In modern spectroscopy, different types of monochromators are employed. Figure 3.5 sketches the setup for a *double monochromator*. In this case, two gratings, three slits, and three mirrors are used. As can be observed, the incoming light is dispersed twice by the gratings. Wavelength scanning occurs by synchronous rotation of both gratings. Obviously, as the number of optical components is increased, the transmitted intensity decreases, due to reflection losses. However, the resolution reached by double monochromators is greatly improved compared to that of single monochromators.

3.3 DETECTORS

As mentioned in Section 1.2 (see also Figure 1.2), different kinds of radiation are involved in optical spectroscopy: incident, transmitted, reflected, dispersed, and emitted. All of this radiation needs to be detected. The great variety of phenomena that are studied by optical spectroscopy means that the radiation involved is spread over a large spectral range, and cannot be monitored by using only one type of detector. As a consequence, much work has been carried out to develop detectors that cover the different spectral ranges required by the optical techniques and the materials studied.

3.3.1 Basic Parameters

When choosing the appropriate detector for an experiment, it is important to look at its *basic parameters*. As we will see, some of these parameters are defined with respect to *noise*. Even in the absence of incident light, detectors generate output signals that are usually randomly distributed in intensity and time. These signals are denoted by noise. The basic parameters of a detector are as follows:

(i) *The spectral operation range.* Common detectors generate an electrical signal (a current or a voltage) that is somehow proportional to the intensity of the beam to be measured. In most detectors, the relationship between the incident intensity and the electrical response is strongly dependent on the photon energy of the incident beam (the incident wavelength). Therefore, depending on the spectral working range, specific detectors must be used.

(ii) *Responsivity.* This is defined as the ratio between the output electrical signal (a current or a voltage) and the incident power. *Responsivity* is usually denoted by R, and is given by

$$R = \frac{V_D}{P} \quad \text{or} \quad R = \frac{I_D}{P} \tag{3.2}$$

where V_D and I_D are the voltage and intensity, respectively, and P is the power of the incident beam. The responsivity is a magnitude that is strongly dependent on the particular wavelength of the incident radiation. As a consequence, the *spectral responsivity at wavelength* λ, R_λ, is usually employed to describe the responsivity of the detector at this input wavelength.

(iii) *The time constant (τ).* Let us suppose that the intensity of the light reaching a detector changes from 0 to I_0 in a very fast (stepwise) way (see Figure 3.6). In an ideal detector, the electrical signal should reproduce the time dependence of the incident intensity. This does not happen in real detectors. Figure 3.6 shows the typical temporal response of a real detector. The output signal increases with time until it reaches a steady value, which is the value corresponding to the incident intensity I_0. The time constant, τ, is defined as the time at which the output signal is 63% (a factor of $(1 - 1/e)$) of the steady output signal (V_0). Detectors with a short time constant are required in the study of fast phenomena.

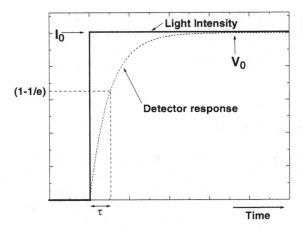

Figure 3.6   The temporal response of a detector after a sudden illumination.

(iv) *The noise equivalent power (NEP)*. This is defined as the incident power of light which produces an output signal equal to the noise of the detector. The *NEP*, which is usually denoted by P_N, is strongly dependent on the particular type of detector and also on its geometric characteristics. For instance, it is well known that P_N grows with the active area of the detector, as the detector noise also increases with this area.

(v) *The detectivity*. This is defined as the inverse of P_N. The detectivity is usually denoted by D, and it tends to be given in W^{-1}.

(vi) *The specific detectivity*. This is an important parameter, which is usually employed to compare different detectors (with different areas and working frequencies). The specific detectivity is denoted by D^*, and is given by

$$D^* = D\sqrt{A \times B} \tag{3.3}$$

where D is the detectivity, A is the area of the detector, and B is the working frequency bandwidth of the detection system (i.e., the detector plus the electronics). Usually, the values of D^* are given for a bandwidth of 300 Hz. D^* is usually be given in units of cm $Hz^{1/2}$ W^{-1}.

3.3.2 Types of Detectors

In spite of the large number of detectors, they can be classified into just two groups: *thermal detectors* and *photoelectric detectors*. We will now discuss the general characteristics of these two types of detectors.

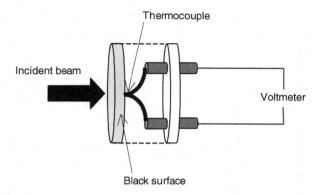

Figure 3.7 A schematic diagram of a conventional thermopile.

Thermal detectors

In these detectors, the light to be measured induces a temperature increase in a given material. This increment is proportional to the intensity of the incident beam. After the corresponding calibration, the intensity of the incident light can be determined by monitoring a temperature-dependent physical magnitude of the material. Although there is a great variety of thermal detectors, in this chapter we will focus our attention on (a) *thermopiles* and (b) *piroelectric* detectors, which are frequently used nowadays in optical spectroscopy laboratories.

(a) *Thermopiles*. Figure 3.7 shows a schematic diagram of a conventional *thermopile*. When the black surface is illuminated, a temperature increment is produced. This temperature increment is monitored by a *thermocouple*, which is attached to the black surface. A thermocouple is a junction between two different metallic wires. As a result of the so-called *Seebeck effect*, any temperature increment in this metallic junction induces a voltage difference. This induced voltage is proportional to the temperature increment and, therefore, somehow proportional to the power of the incident light reaching the black surface. The main advantage of a thermopile is that its responsivity is almost independent of the incident wavelength, since the absorbance of a black surface is almost wavelength independent. This is evident in Figure 3.8, which shows the typical wavelength dependence for the responsivity of a thermopile. Another interesting feature of thermopiles is that they can be used to measure light beams of high intensity, because of the typically high threshold damage of the black surfaces used. On the other hand, since the working procedure of a thermopile involves thermal effects, the main disadvantage of a thermopile is that its time constant is relatively high (in the order of tens of milliseconds). The basic parameters of a thermopile are listed in Table 3.1.

(b) *Piroelectric detectors*. These detectors are based on the temperature dependence of the electric polarization in ferroelectric materials. Figure 3.9 shows the typical

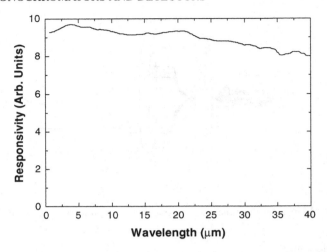

Figure 3.8 The wavelength dependence for the responsivity of a typical thermopile.

temperature dependence of the spontaneous polarization, P, in a ferroelectric material. In piroelectric detectors, the light beam to be measured is focused directly onto the ferroelectric material, or onto a black surface that is in thermal contact with it. For temperatures below the critical temperature ($T < T_C$), the ferroelectric crystal shows spontaneous polarization ($P \neq 0$), which decreases as the temperature increases. Thus, the increment in the crystal temperature caused by the absorption of incident light modifies the value of the spontaneous polarization. If electrodes are applied to the ferroelectric material, then the temporal variation of the temperature (dT/dt) induces an electric current, I, which is given by

$$I = p(T)A\frac{dT}{dt} \tag{3.4}$$

where $p(T)$ is the piroelectric coefficient at each temperature and A is the detector area. As a consequence, in a piroelectric detector the induced current depends

Table 3.1 The basic parameters of different types of detector

Detector	Spectral range (μm)	Time constant	D^* (cm Hz$^{1/2}$ W^{-1})
Thermopile	0.1–40	20 ms	$\approx 1 \times 10^8$
Piroelectric	0.1–40	10 ns – 100 ps	$\approx 1 \times 10^8$
Photoconduction detectors	1–20	$\approx \mu$s	$1 \times 10^9 - 2 \times 10^{11}$
Photodiodes	0.8–4	ns	$1 \times 10^{10} - 2 \times 10^{12}$
Avalanche photodiodes	0.8–2	10 ps – 1 ns	$1 \times 10^{10} - 2 \times 10^{12}$
Photomultiplier	0.1–1.5	0.5–5 ns	

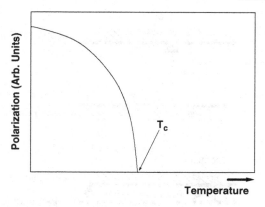

Figure 3.9   Spontaneous polarization versus temperature in a piroelectric detector.

on the rate at which the temperature changes, rather than on its steady value, as happens in the case of thermopiles. Furthermore, as for the case of thermopiles, since a black surface is used to absorb the light beam to be measured, piroelectric detectors work over a very wide spectral range. This is illustrated in Figure 3.10, where the spectral dependence of the responsivity corresponding to a typical piro-electric detector is shown. The basic parameters of piroelectric detectors are listed in Table 3.1. As can be observed, they show very similar detectivities (D*) to those of thermopiles. On the other hand, the time constants of piroelectric detectors are several orders of magnitude shorter than the typical time constants of thermopiles (the time constants of modern piroelectric detectors can be as short as 100 ps).

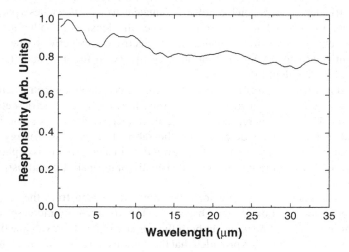

Figure 3.10   The wavelength dependence for the responsivity of a typical piroelectric detector.

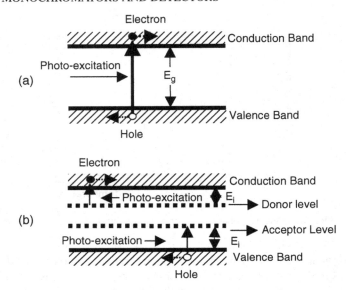

Figure 3.11   Photo-excitation mechanisms in (a) intrinsic and (b) extrinsic photoelectric detectors.

Photoelectric detectors

Photoelectric detectors take advantage of the change in a physical magnitude that occurs as a result of a quantum effect induced by the absorption of photons. The continuous progress in semiconductor technology has led to an ever-increasing use of photoelectric detectors. When light of an adequate wavelength reaches a photoelectric detector and is absorbed, the carrier density (i.e., the density of conduction electrons and holes) changes, causing a change in the electrical conductivity (the resistance) of the semiconductor. After calibration, this change in the conductivity gives the intensity of the incident light.

Depending on the nature of the semiconductor material, photoelectric detectors can be classified as *intrinsic* or *extrinsic* detectors. Intrinsic photoelectric detectors are pure semiconductors, whereas in extrinsic photoelectric detectors some impurities are added to the semiconductor during the fabrication process. Energy diagrams showing the processes activated by *photo-excitation* in these two kinds of photoelectric detectors are shown in Figures 3.11(a) and 3.11(b) for intrinsic and extrinsic detectors, respectively.

In intrinsic photoelectric detectors, electrons are excited from the valence band to the conduction band by photon absorption. The conductivity increases due to the increment in the carrier densities in both the conduction and the valence bands. The excitation process is possible provided that the photon energy of the incident radiation is greater than the energy gap of the semiconductor.

On the other hand, in extrinsic detectors, electrons or holes are created by incident radiation with photons of energy much lower than the energy gap. As can be observed from Figure 3.11(b), the inclusion of impurities leads to donor and/or acceptor energy levels within the semiconductor gap. Thus, the energy separation between these impurity levels and the valence/conduction bands is lower than the energy gap.

The main limitation of photoelectric detectors is the noise caused by thermal excitation of the carriers from the valence band or from the impurity levels. If there is a large *dark current* (a current generated by the detector in the absence of incident light), the sensitivity of the photoelectric detector becomes poor (only very intense beams will induce an appreciable change in the detector conductivity). In order to reduce the dark current, photoelectric detectors are usually cooled during operation.

There are two classes of photoelectric detectors: *photoconduction detectors* and *photodiodes*.

Photoconduction detectors

Figure 3.12 shows the operational scheme of a photoconduction detector. The incident light creates an electrical current and this is measured by a voltage signal, which is proportional to the light intensity. This proportional relation is provided by the fact that, in most photoconduction detectors, the density of carriers in the steady state is proportional to the number of absorbed photons per unit of time; that is, proportional to the incident power.

Figure 3.13 shows the value of the specific detectivity, D^*, for some photoconduction detectors as a function of the incident wavelength. For the sake of comparison, the specific detectivities of a typical thermopile and of a typical piroelectric detector are also shown. As can be observed, photoconduction detectors present higher detectivities (almost two orders of magnitude) than thermal detectors. The main disadvantage of photoconduction detectors is the strong wavelength dependence of their specific

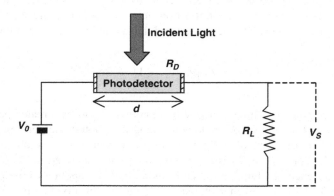

Figure 3.12   The Operational scheme of a photoconduction detector.

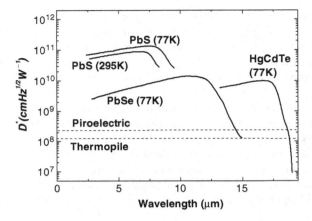

Figure 3.13 The spectral dependence of the specific detectivity for several photoconduction detectors. The values corresponding to a typical thermopile and to a typical piroelectric detector are also shown.

detectivity. In addition, photoconduction detectors cannot be used for the detection of visible radiation, since their detectivity decreases drastically for wavelengths lower than 1 μm.

Photodiodes

Another well-known type of photoelectric detector is the photodiode. Photodiodes can be classified into two types: *p–n photodiodes* and *avalanche photodiodes.*

(a) The p–n photodiodes consist of two doped semiconductors: the p semiconductor is doped in such a way that there is an excess of holes, whereas in the n semiconductor doping leads to an excess of conduction electrons. In an illuminated p–n junction, the relationship between the current and the applied voltage is given by

$$I = I_R \left[e^{eV/kT} - 1 \right] - 2e\eta \frac{P_{\text{opt}}}{h\nu} \qquad (3.5)$$

where η is the quantum efficiency (the number of carriers created per unit time divided by the number of photons reaching the photodiode per unit time), P_{opt} is the power of the incident light (the optical power), $h\nu$ is the photon energy, e is the electron charge, I_R is the electrical current generated in the p–n junction in the absence of illumination, and V is the voltage applied to the photodiode. Figure 3.14 shows the characteristic I–V curves for a typical p–n Si diode for different illumination powers. As can be observed, the I–V curves are strongly dependent on the illumination power, in accordance with Equation (3.5).

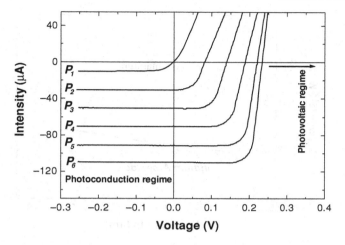

Figure 3.14 Characteristic I–V curves for a typical p–n Si photodiode for different illumination powers ($P_1 < P_2 < P_3 < P_4 < P_5 < P_6$). The two operational regimes of a photodiode detector are also indicated.

Figure 3.14 also allows us to analyze the two operational regimes of a p–n photodiode. In the first regime, the applied voltage is negative, so that expression (3.5) can be written, in a first order approximation, as

$$I \approx -I_R - 2e\eta \frac{P_{opt}}{h\nu}$$

This indicates that the intensity signal increases linearly with the incident power. In this case, for which the applied voltage is negative, it is said that the photodiode is working in the *photoconduction regime*.

Photodiodes can be also used in the *photovoltaic regime*. In this case, the photodiode operates as an open circuit, so that $I = 0$ and Equation (3.5) yields:

$$V = \frac{kT}{e} \ln \left[\frac{2e\eta}{I_R} \times \frac{P_{opt}}{h\nu} + 1 \right] \tag{3.6}$$

In this regime, the signal voltage is not proportional to the light power, but follows a logarithmic trend. In addition, in this configuration the time constant of the detector can be as short as a few nanoseconds.

Figure 3.15 shows the spectral dependence of the *specific detectivity*, D^*, reported for germanium and indium arsenide *photodiodes*, respectively. The main properties of these detectors are also summarized in Table 3.1 for comparison.

(b) An *avalanche photodiode* is a p–n junction with a high doping level. This high doping leads to a strong curvature of both the valence and the conduction bands in the proximity of the p–n junction. When a highly doped photodiode is inversely polarized (i.e., a negative voltage is applied), the carriers created by illumination

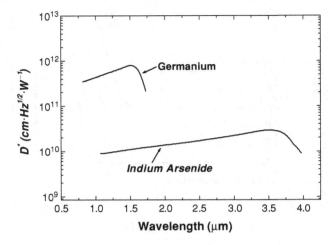

Figure 3.15 The specific sensitivities reported for germanium and indium arsenide photodiodes.

are strongly accelerated by the electric field induced in the junction. The carriers then acquire enough kinetic energy to create new electron–hole pairs by elastic collisions. These processes can take place several times, so that the absorption of one photon can generate several carriers. The number of carriers generated per absorbed photon is known as the *multiplication factor*. Obviously, the multiplication factor increases with the acceleration voltage (the applied voltage). Figure 3.16 shows the multiplication factor of a typical avalanche photodiode as

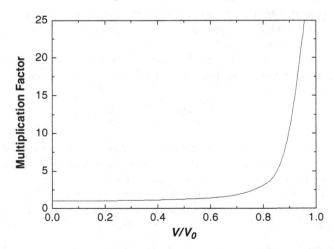

Figure 3.16 The multiplication factor of a typical avalanche photodiode as a function of the applied voltage.

a function of the applied voltage V (V_0 being the breakdown voltage of the avalanche photodiode). The main advantage of avalanche photodiodes over p–n photodiodes is their time constant. In avalanche photodiodes, the carriers are subjected to a strong electric field that induces a large acceleration, leading to very short transit times. The time constants of avalanche photodiodes usually range from tens of picoseconds up to one nanosecond (several orders of magnitude lower than in the case of p–n photodiodes). Their main disadvantage is their small detection area. This is due to technical limitations; it is not easy to fabricate large highly doped p–n junctions with enough spatial homogeneity.

3.4 THE PHOTOMULTIPLIER

Although the *photomultiplier* can be considered to be a photoelectric detector, in this chapter we will describe it in a separate section, because of the special relevance of photomultipliers in the field of optical spectroscopy. This detector is more complicated and expensive than those described in previous sections. Nevertheless, photomultipliers are probably the most common detectors used in optical spectroscopy experiments. The reason for this is their high sensitivity and stability.

3.4.1 The Working Principles of a Photomultiplier

A schematic drawing of a photomultiplier is shown in Figure 3.17. A photomultiplier consists of a *photocathode*, a chain of *dynodes*, and a *collector* (anode). The light to be detected illuminates the photocathode (the active area of the photomultiplier), which generates electrons due to the absorption of incident photons. These electrons are accelerated and amplified by the dynodes and, finally, they arrive at the anode, where are monitored as an induced current.

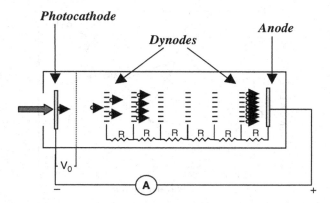

Figure 3.17   A schematic representation of a photomultiplier.

The photocathode consists of a material with a very low work function (defined as the energy required by an electron to get out of the material). Multi-alkali compounds are widely used, although some semiconductors, such as GaAs and InGaAs, are also employed. When semiconductor photocathodes are used, the response of the photomultiplier in the deep-red and near infrared spectral regions is improved. The quantum efficiency of a given photocathode is defined as the number of electrons released per incident photon. A typical magnitude sometimes used to describe the response of a photocathode is its responsivity, R, which was defined in the previous section as the light-induced current divided by the power of the incident beam. The responsivity of a photocathode is strongly related to its quantum efficiency, η. Taking into account that the charge generated per incident photon is $\eta \times e$ (e being the electronic charge), we can easily obtain that

$$R = \frac{e \times \eta \times \lambda}{hc} \tag{3.7}$$

where λ is the wavelength of the incident light, h is Planck's constant, and c is the speed of light. Recall that the responsivity is a function of the incident wavelength ($R \equiv R_\lambda$). In Figure 3.18, this is shown by the dependence of η on λ.

Figure 3.18 shows the wavelength dependence of the quantum efficiency of several photocathodes. As can be observed, in all the cases the quantum efficiency is below 30 % and it drops down to zero in the near infrared. Great efforts are now being made to develop new photocathodes with an extended response in the infrared. Nowadays, it is possible to find commercial photomultipliers with a nonvanishing response up to 1.5 μm. Nevertheless, as a general rule, the broader the spectral range, the lower is the quantum efficiency of the photocathode.

At this point it should be mentioned that, as occurs for other detectors, even in the absence of illumination, electrons are emitted by the photocathode due to

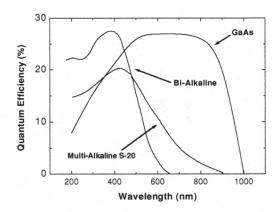

Figure 3.18   Quantum efficiency versus wavelength of several photocathodes.

thermal activation. As was mentioned before, photocathodes are made from materials with low work functions, so that thermal energy can exceed the energy required for electron emission. As we will see later, this is one of the main sources of noise in photomultipliers. Fortunately, the dark emission from photocathodes can be efficiently reduced by appropriate cooling.

Once electrons have been emitted by the photocathode, they are accelerated by an applied voltage induced between the photocathode and the first dynode (V_0 in Figure 3.17). The dynodes are made of CsSb, which has a high coefficient for secondary electron emission. Thus, when an electron emitted by the photocathode reaches the first dynode, several electrons are emitted from it. The amplification factor is given by the *coefficient of secondary emission*, δ. This coefficient is defined as the number of electrons emitted by the dynode per incident electron. Consequently, after passing the first dynode, the number of electrons is multiplied by a factor of δ with respect to the number of electrons emitted by the photocathode. The electrons emitted by this first dynode are then accelerated to a second dynode, where a new multiplication process takes place, and so on. The *gain of the photomultiplier*, G, will depend on the number of dynodes, n, and on the secondary emission coefficient, δ, so that

$$G = \delta^n \tag{3.8}$$

Taking a typical value of $\delta \cong 5$ and considering 10 dynodes, Equation (3.8) gives a gain of $G = 5^{10}$ (which is of the order of 10^7). The particular value of δ depends, of course, on the dynode material and on the voltage applied between dynodes. Similarly to the case of the photocathode, the responsivity of a photomultiplier, R^{PM}, is defined as the current induced in the anode divided by the power of the light reaching the photocathode. Thus, it is very simple to show that

$$R^{PM} = \frac{e \times \eta \times \lambda \times G}{hc} \tag{3.9}$$

where, again, λ is the wavelength of incident light and G is the gain of the photomultiplier given by Equation (3.8).

The way in which the voltage is applied between dynodes depends on whether the photomultiplier is to be used in the continuous wave or in the pulsed regime. For continuous wave measurements, the voltage is applied following the scheme of Figure 3.17. In this configuration, there is a fixed resistance R between any two consecutive dynodes. As a consequence, the voltage between consecutive dynodes is the same for all of the dynodes and it is controlled by applying a constant current I, so that a constant voltage equal to IR is induced between the dynodes. This acceleration voltage between the dynodes is usually around 150 V, using resistances of the order of 100 kΩ.

When the light to be detected consists of a train of pulses, very high peak currents are generated inside the photomultiplier tube and a different arrangement is required (see Figure 3.19). In this case some capacitors, C, are included between the last

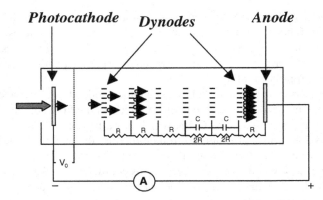

Figure 3.19   A schematic drawing of a photomultiplier specially designed for the measurement of pulsed radiation.

dynodes. The function of these capacitors is to avoid any change in the acceleration voltage induced by the high electron currents circulating inside the photomultiplier. For pulse currents of 10 mA, the values of the capacitances used are of the order of 20 nF.

Once the electrons have been accelerated and multiplied, they reach the anode. The electrons arriving at the anode produce an electrical current. This current can be measured directly, or indirectly by monitoring the voltage increment induced in a given load resistor, R_L. This load resistor is critical, as it determines the time constant of the photomultiplier. A typical time constant for a photomultiplier is 2 ns, although an adequate choice of the load resistor and anode material could lead to time constants as low as 0.5 ns.

When time-dependent signals are to be measured by a photomultiplier, the time sensitivity is usually limited by the inhomogeneous *transit time*. The transit time is the time taken by electrons generated in the cathode to arrive at the anode. If all of the emitted electrons had the same transit time, then the current induced in the anode would display the same time dependence as the incoming light, but delayed in time. However, not all of the electrons have the same transit time. This produces some uncertainty in the time taken by electrons to arrive at the anode. There are two main causes of this dispersion:

(i) The electrons (photoelectrons and those generated by dynodes) follow different trajectories to reach the anode.

(ii) The photoelectrons are not emitted with the same velocity, neither from the cathode nor from the dynodes.

Figure 3.20 shows the effect of the transit time dispersion on the measurement of an ideal light pulse. Since photoelectrons spend some time traveling from the photocathode to the anode (transit time), the photomultiplier signal is delayed in time with respect to the incident pulse. Furthermore, due to the transit time dispersion, the

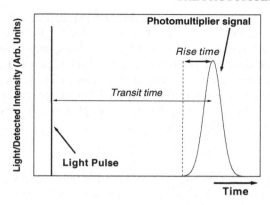

Figure 3.20 The time dependence of an ideal sharp light pulse (solid line) and the temporal evolution of the corresponding electrical signal generated by the photomultiplier. The rise time is indicated.

generated signal does not follow the temporal shape of the incident light. As can be observed in Figure 3.20, the detected signal is broader than the incident pulse. We can now introduce another important parameter in photomultipliers, the *rise time*. The rise time of a photomultiplier is the time between the start of the signal pulse and the signal maximum (provided that the signal is generated by an 'ideal sharp pulse'). This is indicated in Figure 3.20. The transit time dispersion, and consequently the rise time, are the main limitations for the use of photomultipliers in the sub-nanosecond range.

3.4.2 Noise in Photomultipliers

As was mentioned before, noise is a term used to describe any random output signal that has no relationship with the incoming signal (the incoming light). In photomultipliers, noise can be classified, depending on its origin, into three types: *dark current*, *shot* noise, and Johnson noise. The differences between these three classes are explained next:

Dark current noise

Even in the absence of illumination (darkness) some electrons, excited by thermal energy, are emitted from the photocathode. Since photocathodes are materials with low working functions, the thermal energy can be high enough to induce the emission of electrons. These emitted electrons give rise to what is known as the dark current or, sometimes, the *thermo-ionic current*. The dark current varies randomly with time, so that it is considered as noise. It has been experimentally determined that the thermo-ionic current, I_t, due to photoelectrons emitted by a photocathode in the absence of illumination is given by

$$I_t = aAT^2 e^{-e\phi/kT} \tag{3.10}$$

where k is Boltzmann's constant, a is a constant that depends on the photocathode material (for pure metals, $a = 1.2 \times 10^6$ m^{-2} K^{-2} A), A is the area of the photocathode, T is photocathode temperature, and ϕ is the corresponding work function. It is clear that the dark current depends on the photocathode material through the value of the constant a and of the work function, ϕ. Nevertheless, for a given material, the dark current can be minimized by reducing the photocathode area and also by cooling down the photomultiplier. In fact, most photomultipliers are cooled by a *Peltier* cooler and/or a closed water circuit.

EXAMPLE 3.1   *(a) Calculate the dark current intensity at room temperature (T = 300 K) of a metallic photocathode with the following characteristics: area = 10 cm^2, eϕ = 1.25 eV.*

(b) Now calculate the dark current intensity generated in the anode (output dark current) if the photomultiplier has 10 dynodes, each of which has a secondary emission coefficient of $\delta = 4$.

(c) Estimate the reduction in the dark current intensity reached when the photocathode is cooled down to 5 °C.

(a) The dark current at room temperature ($kT = 0.025$ eV) is given by expression (3.10):

$$I_t \, (T = 300 \, \text{K}) = 1.2 \times 10^6 \times 10^{-3} \times (300)^2 \times e^{-1.25/0.025} \approx 2 \times 10^{-14} \text{A}$$

This is the photocathode dark current.

(b) In photomultipliers, this current will be amplified by the dynodes. The amplification factor is given by expression (3.8), so that

$$G = 4^{10} \approx 10^6$$

Therefore, the output dark current (I_t^{Out}) is given by

$$I_t^{Out} = I_t \times G = 2 \times 10^{-14} \times 10^6 = 2 \times 10^{-8} \text{A} = 20 \, \text{nA}$$

(c) If the photomultiplier (and hence the photocathode) is cooled down to 5 °C = 278 K, the photocathode dark current will be given by

$$I_t \, (T = 278 \, \text{K}) = 1.2 \times 10^6 \times 10^{-3} \times (278)^2$$
$$\times e^{-1.25/0.024} \approx 2 \times 10^{-15} \text{A}$$

so that the photocathode dark current has been reduced by one order of magnitude. Obviously, the output dark current would be also reduced by one order of magnitude.

Shot noise

This noise source is associated with the discrete nature of the electric current. When a certain current i is induced or generated in the photocathode, there is some uncertainty in the current, which arises from the quantum properties of electrons. It has previously been demonstrated that the fluctuations in any electrical current with a frequency between f and $f + \Delta f$ are given by

$$\Delta i = \sqrt{2ie\Delta f} \tag{3.11}$$

In the particular case of a photocathode, this fluctuation affects both the dark current (i_t) as well as the illumination induced current (i_{lum}). In the absence of illumination, the only current generated in the photocathode is the dark current, and so the shot noise associated with it is Δi_t. If the light-induced current, i_{lum}, is smaller than the shot noise associated with the dark signal (Δi_t), then it will be not possible to distinguish any light-induced current. In these conditions, the incident light cannot be detected by the photomultiplier, as it is not possible to separate the noise and the signal. As a consequence, the shot noise associated with the dark current determines the minimum intensity that can be detected by a particular photomultiplier (or by a particular photocathode). This is clearly shown in the next example.

EXAMPLE 3.2   (a) Calculate the minimum light power detectable by a photocathode whose quantum efficiency is 0.2 at 400 nm and with the following characteristics: area = 10 cm², eφ = 1.25 eV. Assume an incident wavelength of 400 nm and a bandpass width of 1 Hz. Estimate the minimum intensity detectable by the photocathode if it is cooled down to 5 °C.
    (b) Now calculate the minimum light power that can be measured with a photomultiplier made up with the previous photocathode and with 10 dynodes, each of which has a secondary emission coefficient of δ = 4

(a) The photocathode responsivity is given by Equation (3.7). Thus introducing our data in MKS units, we obtain:

$$R = \frac{e \times \eta \times \lambda}{hc} = \frac{1.6 \times 10^{-19} \times 0.2 \times 0.4 \times 10^{-6}}{6.62 \times 10^{-34} \times 3 \times 10^{8}} = 0.06 \, \text{A W}^{-1}$$

As we found in Example 3.1, the dark current of this photocathode when operating at room temperature is $I_t \, (T = 300 \, \text{K}) \approx 2 \times 10^{-14}$ A. The minimum current that can be measured is equal to the current dispersion caused by shot noise over the dark current, so that

$$I_{\text{min}} = \Delta I_T = \sqrt{2I_T e\Delta f} = \sqrt{2 \times 2 \times 10^{-14} \times 1.6 \times 10^{-19} \times 1}$$
$$= 8 \times 10^{-17} \text{A}$$

As a consequence, the minimum power that can be detected by the photocathode will be given by

$$W_{min} = \frac{I_{min}}{R} = \frac{8 \times 10^{-17}}{0.06} = 1.3 \times 10^{-15} \text{ W}.$$

This corresponds to a photon flux equal to

$$\phi_{min} = \frac{W_{min}}{hc/\lambda} = \frac{1.3 \times 10^{-15}}{6.62 \times 10^{-34} \times 3 \times 10^8 / 0.4 \times 10^6} = 2620 \text{ photons per second}$$

If the photocathode is cooled down to 5 °C, the dark current is reduced from 2×10^{-14} down to 2×10^{-15} A. As a result, the minimum current detectable at 5 °C is given by

$$I_{min} (5 °C) = \Delta I_T (5 °C) = \sqrt{2 I_T (5 °C) e \Delta f}$$
$$= \sqrt{2 \times 2 \times 10^{-15} \times 1.6 \times 10^{-19} \times 1} = 2 \times 10^{-17} \text{ A}$$

Following the previous arguments, it can be determined that the minimum photon flux that can be detected by the photocathode at 5 °C is 655 photons per second, which is four times lower than the minimum photon flux detectable at room temperature.

(b) When we are dealing with the minimum intensity that can be detected by the photomultiplier, the responsivity to be considered is R^{PM}, which depends on the particular gain of the photomultiplier studied ($G \approx 10^6$, as calculated in Example 3.1). Therefore, R^{PM} is given by

$$R^{PM} = \frac{e \times \eta \times \lambda \times G}{hc}$$
$$= \frac{1.6 \times 10^{-19} \times 0.2 \times 0.4 \times 10^{-6} \times 1 \times 10^6}{6.62 \times 10^{-34} \times 3 \times 10^8} = 6 \times 10^4 \text{ A W}^{-1}$$

Additionally, the photocathode dark current will be amplified by a factor G before arriving at the anode. Therefore, the minimum intensity that can be detected at the anode, at room temperature, is given by

$$I_{min}^{PM} = \Delta I_T^{PM} = \sqrt{2 I_T G e \Delta f}$$
$$= \sqrt{2 \times 2 \times 10^{-14} \times 1 \times 10^6 \times 1.6 \times 10^{-19} \times 1} = 8 \times 10^{-14} \text{ A}$$

so that

$$W_{min}^{PM} = \frac{I_{min}^{PM}}{R^{PM}} = \frac{8 \times 10^{-14}}{6 \times 10^4} = 1.3 \times 10^{-14} \text{ W}$$

and consequently

$$\phi_{min}^{PM} = \frac{W_{min}^{PM}}{E_{ph}} = \frac{1.3 \times 10^{-14}}{6.62 \times 10^{-34} \times 3 \times 10^8 / 0.4 \times 10^6} \approx 2 \text{ photons per second}$$

This is indeed a really low photon flux.

Johnson noise

This noise is due to the thermal motion of the carriers (electrons) in the different resistors used in the photomultiplier. In general, the signal uncertainty caused by this source of noise is much lower than those generated by both dark noise and shot noise.

3.5 OPTIMIZATION OF THE SIGNAL-TO-NOISE RATIO

In many optical spectroscopy experiments, the intensity of the radiation involved is very low. In these conditions, the noise can be as intense as the signal. The *signal-to-noise ratio* is the quantity that characterizes the quality of the measured signal. There are several methods that are especially devoted to increasing this ratio. Some of them deal with the measurement procedure, whereas others involve the use of some specific electronic devices to treat the measured signal. The most usual methods employed for the increment in the signal-to-noise ratio are listed next.

3.5.1 The Averaging Procedure

Of course, the simplest way to optimize the signal-to-noise ratio is to perform consecutive measurements and then calculate the average signal. In this way, the noise will be reduced as it is randomly distributed in time. The signal-to-noise ratio increases with the number of averaged measurements. If we denote by N the number of signals averaged, it is postulated from basic principles of statistics that the signal-to-noise ratio increases as $\sqrt{N}$. This method represents an effective way of enhancing the signal-to-noise ratio, but it does implicate the repetition of a given measurement a large number of times and, consequently, the time that should be employed for the measurement is increased.

3.5.2 The Lock-In Amplifier

As was noted before, several techniques based on the electronic analysis of signals have been developed to increase the signal-to-noise ratio without requiring large measurement times. In this section, we will focus our attention on the lock-in amplifier and in the next section on photon counter systems.

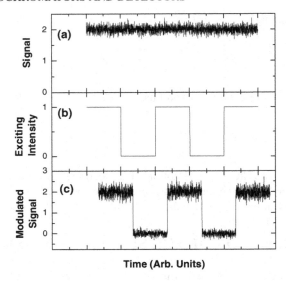

Time (Arb. Units)

Figure 3.21 (a) The signal (noise plus sample signals) obtained when a material is illuminated by a continuous light beam. (b) The modulation pattern obtained for the excitation intensity by means of a mechanical chopper. (c) The signal (noise plus sample signals) obtained when a material is illuminated by mechanically modulated light beam. Note the presence of a certain amount of phase mismatch in respect to the excitation signal.

When a material is illuminated with a beam of constant intensity (continuous illumination) the radiation signals coming from it (fluorescence, scattered, reflected, and transmitted light) will be also of constant intensity in time, except for the possible contribution coming from the different noise sources. The fraction of the signal due to noise is, in general, randomly distributed in time. On the contrary, the sample signal is independent of time, as it is produced by the exciting beam (whose intensity is assumed to be time independent). This has been illustrated in Figure 3.21(a), which represents the total signal (noise plus sample signal) obtained under continuous illumination (excitation). If the exciting light is modulated, the intensity of the sample signal will be also modulated, as it is caused by the exciting beam. Continuous light sources are usually modulated by a *mechanical chopper*, which is a rotating wheel with evenly spaced slots on it. The light chopper is usually equipped with a photocell and suitable electronics to generate an electrical signal modulated at the chopping frequency. This electrical signal is used as the reference signal for the instruments used in later signal processing. Figure 3.21(c) shows the signal intensity (fluorescence, scattered, reflected, or transmitted intensity) obtained under excitation with a mechanically modulated excitation beam, whose intensity is displayed in Figure 3.21(b). From Figure 3.21(c), it is clear that noise will be present even when the pump light is blocked by the chopper. From this figure, it is also clear that the noise-induced signal has a large range of frequencies (it is randomly distributed in time), whereas the sample signal intensity has the same frequency as the pump beam (modulating frequency). Finally, it is important to note that a certain amount of *phase mismatch* could appear between the exciting beam and the signal intensity. This phase mismatch could be caused by various physical processes, which are beyond the scope of this chapter. The possible existence of phase mismatch can be also noted in Figure 3.21(c) (note that the modulated signal is displaced in time with respect to the exciting intensity).

A *lock-in amplifier* is, first of all, an amplifier that it is tunable to a selected frequency, so that it can selectively amplify signals modulated at a given frequency. For signal measurement, the lock-in amplifier is tuned to the pump chopping frequency by using the reference signal provided by the photocell. As a consequence, only electrical signals with the same frequency as that of the reference signal (i.e., the sample signal) will be amplified, whereas signals with other frequencies (noise) will not be amplified. In addition, the phase between the reference signal provided by the chopper and the measured signal can be varied by the lock-in amplifier in a controlled way, in order to obtain maximum amplification. This situation is achieved when the phase artificially induced in the reference signal is equal to the phase mismatch between the exciting beam and the measured signal. Therefore, lock-in amplifiers are sensitive to phase mismatches between the excitation and the measured signals, this being a valuable characteristic.

Most common lock-in amplifiers can be operated at frequencies ranging from a few Hz up to 100 kHz. This fact is important in analyzing the temporal evolution of optical signals; for example, fluorescence decay time measurements. Although this particular application of lock-in amplifiers is beyond the scope of this section, it is instructive to mention that this can be done by tuning the relative phase (the time delay) between the signal intensity and the reference signal provided by the chopper.

3.5.3 The Photon Counter

The *photon counter* is used in the processing of weak signals generated from a photomultiplier. The photomultiplier output signal generally consists of a series of discrete peaks with different intensities (see Figure 3.22). The reason for this is that,

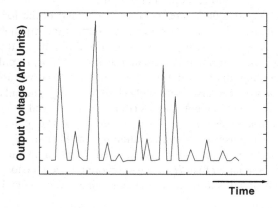

Figure 3.22   Output signal versus time generated by a typical photomultiplier.

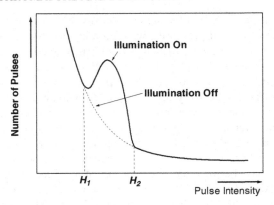

Figure 3.23   The number of pulses generated from a photomultiplier as a function of the pulse intensity (pulse height) for the illumination-on and illumination-off regimes (see the text).

as described earlier in this chapter, photoelectrons (those induced by illumination) and thermally ejected electrons emerge from the photocathode and are amplified by the dynode chain. However, the spikes corresponding to the thermal electrons are, in general, smaller than those due to photoelectrons (i.e., the number of electrons generated at the photocathode by illumination is larger than the number of electrons that are thermally emitted).

For certain photomultipliers (with high quantum efficiencies, short transit times, and short rise times) the number of pulses (spikes) generated in the absence of illumination is strongly related to the intensity of these pulses. This relation is illustrated in Figure 3.23. As can be observed, in the absence of any illumination (in the *illumination-off regime*), the number of generated pulses is only appreciable in the low pulse intensity region. This is, of course, reasonable, as in the illumination-off regime pulses are due to noise, which usually produces low-intensity spikes. If the photomultiplier is now illuminated (the *illumination-on regime*), the number of pulses with intensities within a certain range increases drastically. Since, even in the case of the illumination-on regime, thermal emission is also present, both (illumination-on and illumination-off) curves are practically identical in the low pulse intensity range (a pulse intensity $I < H_1$). However, in a certain intermediate range of pulse intensities ($H_1 < I < H_2$), the two curves are very different, so that the number of pulses measured under illumination is greater than the number of pulses measured in darkness. For very high intensity pulses ($I > H_2$), the two curves are similar, as illumination does not produce pulses with such a high intensity. The pulse intensities that define the range within which both curves (illumination-on and -off) are different are called *discrimination levels* (H_1 and H_2 in Figure 3.23). The photon counter simply counts the number of pulses with heights ranging from H_1 up to H_2. All of the pulses or peaks with intensities lower than H_1 are ignored. At the same time, all of the pulses or peaks with a height larger than H_2 are also ignored. In conclusion, the contribution of noise to the net signal is minimized and, consequently, the signal-to-noise ratio is strongly enhanced.

3.5.4 The Optical Multichannel Analyzer

The use of *optical multichannel analyzers* (OMAs) has increased the signal-to-noise ratio in the measurements of weak absorption and luminescence spectra. The basic operational scheme of an OMA is illustrated in Figure 3.24.

As can be observed, the arrangement of a conventional OMA is quite similar to that of a monochromator. A diffraction grating is used for a spatial separation of the different spectral components of the incoming radiation. The dispersed light is then analyzed by a one-dimensional arrangement of very narrow spaced diodes/detectors (the number of diodes is usually 1024). In this configuration, the intensity associated with each spectral component is recorded by an individual diode. This setup allows for a fast recording of the whole spectrum, as the requirement of grating rotation is avoided. The main advantage of OMAs is that they allow for the measurement of the spectrum in real time. When adequate electronics are used, OMAs provide the possibility of individual averaging or integration of each spectral component. Diode arrays are usually made of Si, so that their spectral response covers a broad wavelength range, typically from 400 nm up to 1100 nm. Typical dimensions for the diodes (pixels) are 25 μm $\times$ 2.5 mm, with a pixel-pixel distance of 25 μm. With fast scanning diode arrays, scan times as low as 13 μs per diode can be achieved. This makes it possibile to record about 70 spectra per second. When OMAs are compared with standard monochromator plus photomultiplier equipment, the sensitivity of Si arrays is usually more than one order of magnitude lower. However, for a typical array of 1024 diodes, a factor of about 1000 is gained in measuring time.

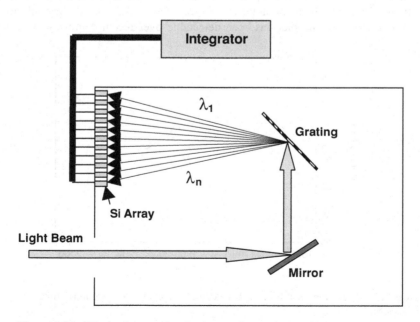

Figure 3.24   The basic operational scheme of an optical multichannel analyzer.

This allows for a high amount of integrations or averaging procedures, leading to a substantial reduction in the signal-to-noise ratio. On the other hand, one disadvantage of OMAs is their spectral resolution, which is lower than that provided by monochromators.

3.6 DETECTION OF PULSES

The measurement of physical phenomena with a temporal dependence is crucial for a full understanding of light–matter interaction processes, and also in optical spectroscopy. As we have shown in Section 1.4, this is especially relevant in the case of time-resolved luminescence. Recall that when a given material is excited by a pulse of light with an adequate wavelength, this incident radiation is absorbed by optical centers, which are excited to higher energy states. Once the excitation pulse has expired, the centers relax to lower energy states through the emission of photons and/or phonons. The time evolution of the emitted intensity provides valuable information about the nature of the physical phenomena involved. If an adequate detector is used, then the signal provided by the detector will display the same time dependence as the luminescence. Nevertheless, the recording of fast signals requires special systems. As in the case of detectors, the particular experimental tool used depends on the timescale over which the phenomena involved are taking place. Consequently, there is no a universal experimental setup that is valid for all of the processes of interest in optical spectroscopy.

The timescale of the physical processes that can be detected optically is decreasing continuously due to the technological advances. Figure 3.25 shows the evolution of the minimum timescale that has been detectable over recent centuries. As can be

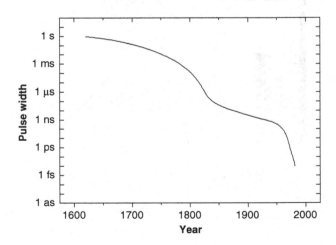

Figure 3.25   The minimum timescale that has been detectable as a function of time over recent centuries.

observed, the time sensitivity has decreased by a factor of more than 10^{12} during the past 300 years. In optical spectroscopy, the majority of the physical phenomena occur in a timescale from milliseconds down to femtoseconds.

The main schemes and experimental devices used for the detection of signals within this time range are briefly described below. In Section 3.7 we will also introduce the instrumentation used to measure optical signals in the picosecond–femtosecond range.

3.6.1 Digital Oscilloscopes

For the measurement of timescales ranging from several milliseconds down to hundreds of picoseconds, the combination of an adequate detector together with a *digital oscilloscope* constitutes the simplest experimental setup. Modern oscilloscopes work with bandwidths as high as 6 GHz and with temporal resolutions as short as 50 ps. Commercial photodetectors with time constants of the order of 100 ps are available. Furthermore, the time constants of avalanche photodiodes are in the tens of picoseconds range. Consequently, an adequate oscilloscope and detector combination will allow us to monitor the time evolution of the optical signal, with timescales typically larger than hundreds of picoseconds. One of the main advantages of this experimental setup is that modern digital oscilloscopes allow for very large integration times, and so the continuous averaging of repetitive signals leads to high signal-to-noise ratios.

3.6.2 The Boxcar Integrator

In some experimental arrangements, *boxcar integrators* are used instead of digital oscilloscopes. The basic operational scheme of a boxcar integrator is shown in Figure 3.26.

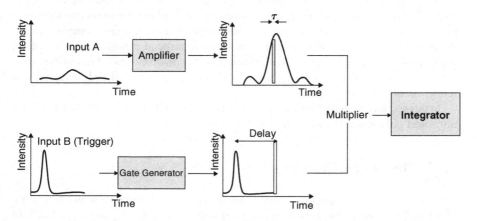

Figure 3.26 The operational scheme of a boxcar integrator.

A boxcar integrator has two inputs. From one of them (Input A), the detected signal is amplified properly. The other input (Input B) acts as a trigger. The trigger signal can be generated by the boxcar integrator itself (an internal trigger), or by the pump source, typically a laser, so that the time at which the sample was excited ($t = 0$) is determined. The signal from Input B is treated by a gate generator, which provides a square signal with a certain delay, t, with respect to the reference pulse (the trigger). This square signal is usually referred to as the *gate* and its corresponding width is denoted by τ. The gate signal is fed into a multiplier circuit together with the amplified signal from Input A. The function of this multiplier circuit is just to multiply both signals, so that only the fraction of signal overlapping in time with the gate is selected. This selected fraction is then launched into an integrator, in which averaging with other subsequent signals is carried out. In this way, the intensity of the signal at time t (with respect to the excitation time) is measured. The delay time can be varied and the integrator is used to give the intensity versus time curve corresponding to the averaged signal. With adequate gate widths, time resolutions in the order of 100 ps can be obtained. More sophisticated boxcar integrators use several gates, properly distributed in time, instead of only one. In this case, each gate amplifies the signal at a given time t with respect to the excitation. Consequently, the time dependence of the signal is obtained by monitoring the amplified signal corresponding to each gate, instead of varying the delay time of a single gate, as happens in the simplest boxcar integrators. These multi-gate boxcar integrators are called *multichannel analyzers*. Obviously, the time resolution of multichannel analyzers increases with the number of gates used for signal amplification.

3.7 ADVANCED TOPICS: THE STREAK CAMERA AND THE AUTOCORRELATOR

Most optical centers show luminescence decay times in the nanoseconds–milliseconds range. However, many other physical processes involved in optical spectroscopy are produced in the picoseconds–femtoseconds range, and much more complicated instrumentation becomes necessary. For instance, interband luminescence in solids, which is of particular interest in semiconductors, can involve decay times in the range of picoseconds. Pulses generated from solid state lasers have already reached this femtosecond domain.

In this section, we will briefly describe the two main techniques devoted to detecting ultrashort pulses: the *streak camera* and the *autocorrelator*.

3.7.1 The Streak Camera

The *streak camera* is especially suitable for the analysis of signals within a timescale of the order of picoseconds. The operational scheme of a *streak camera* is shown in Figure 3.27.

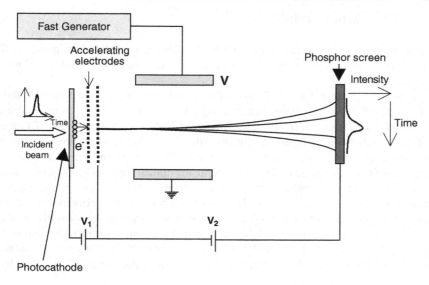

Figure 3.27   The basic operational scheme of a streak camera.

The incident light pulse passes through a slit and its corresponding image is formed on a photocathode, generating the emission of electrons. The number of electrons emitted is proportional to the light intensity arriving to the photocathode at a given instant. These emitted electrons pass thorough a pair of accelerating electrodes, so that they are directed to a *phosphor screen*. Between the photocathode and the phosphor screen, the emitted electrons are deviated by an external electric field generated by two electrodes connected to a fast voltage generator, as shown in Figure 3.27. The high voltage between these electrodes is synchronized with the incident light and it varies very rapidly in time. During this high-speed sweep, electrons emitted at slightly different times from the photocathode are deviated by different electric fields and, consequently, they are deflected at different angles in the vertical direction. Therefore, they arrive at different positions on the phosphor screen, so that the vertical direction on this screen (see Figure 3.27) serves as the time axis. The brightness at a given point on the vertical axis of the screen is proportional to the intensity of the incident light at the corresponding time. In such a way, the temporal profile of the incident pulse is reproduced.

The streak camera provides the possibility of time resolutions better than 1 ps. For weak signals, some integration over the screen signal is required and this leads to a worse time resolution (of the order of 10 ps). The main advantage of the streak camera is that it provides a great amount of information in a short time. On the other hand, its main disadvantages are a high cost and a limited spectral range of detection (500–1100 nm).

3.7.2 The Autocorrelator

The *autocorrelator* is an important tool in the detection of ultrashort pulses in the femtosecond–picosecond range. The basic operational scheme of an autocorrelator is shown in Figure 3.28.

In a first step, the pulse of light to be analyzed is divided into two pulses (A and B in Figure 3.28) of similar intensities by using a beam splitter. In a second step, one of these two pulses (pulse A in our case) is delayed with respect to the other. There are different mechanisms to induce a controlled delay between two synchronized pulses. The time required by a pulse to travel a certain distance is given by L/c, L being the optical path and c the speed of light. In general, the optical path can be expressed as the distance traveled by the pulse (l) multiplied by the refractive index of the propagating medium (n). Any change, in either the traveling distance or the refractive index, will cause a certain delay between pulses. In the schematic drawing shown in Figure 3.28, we have assumed that the pulse delay is induced by a controlled modification in the traveling distance l. One pulse (pulse A in Figure 3.28) travels over a greater distance than the other (pulse B). The actual delay time between the pulses, Δt, is given by

$$\Delta t = c \times \Delta x \qquad (3.12)$$

where Δx is the increment in the distance traveled by pulse A with respect to pulse B. A controlled increment in the traveling distance is achieved by using a moving mirrors system (see Figure 3.28). When the moving mirrors system is translated parallel to the pulse direction by a distance a, then the net traveling distance of pulse A will be increased by an amount of $2a$. In general, the moving mirrors are mounted on a

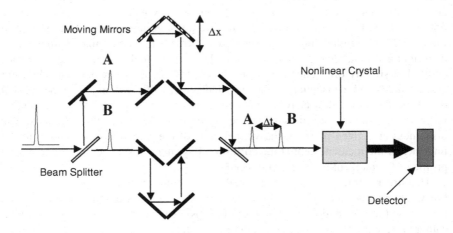

Figure 3.28 The basic operational scheme of an autocorrelator.

piezoelectric controlled translation stage, so that fine adjustment of the mirror can be achieved.

Once this time-controlled delay has been induced, both pulses are spatially over-lapped and focused into a *nonlinear crystal*. If this nonlinear crystal satisfies some geometric conditions (phase-matching conditions, as defined in Section 2.5), then frequency summation between these two pulses will be possible, leading to the gen-eration of light with a wavelength equal to the half-wavelength of the incident pulses (recall that both pulses have the same photon energy).

The intensity of the second harmonic beam depends on a great variety of factors, such as geometric factors (the focused size and polarization of the incident light), the nonlinear crystal temperature, the nonlinear conversion efficiency, and the particular time delay of the interacting pulses. If the time delay is greater than the pulse width, pulses will not overlap in time inside the nonlinear crystal and the intensity of the second harmonic wave will be at a minimum (see Figure 3.29). As the time delay is decreased, some temporal overlap between the two pulses occurs inside the nonlinear crystal, so that the intensity of the second harmonic wave increases. On the other hand, if the time delay is zero, then the intensity of the second order wave will be at maximum. Therefore, by monitoring the intensity of the second harmonic wave as a function of the difference in traveling distance (or moving mirrors displacement) and applying relation (3.17), it is possible to plot the intensity of the second harmonic wave as a function of the time delay. Then, from this measurement, the pulse width and shape can be determined.

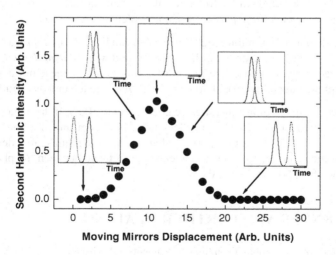

Figure 3.29 The intensity of the second harmonic wave generated in an autocorrelator as a function of the displacement of the moving mirrors system. The insets show the intensity versus time curves for pulses A and B of Figure 3.28 (solid and dashed lines, respectively).

EXERCISES

3.1 A light beam of 21 mW reaches a photoconduction detector with a 1 mm thick active area. The absorption coefficient at 965 nm (the incident wavelength) is 23 cm^{-1}. Calculate the number of carriers created per unit time if the quantum efficiency of the process is 0.13.

3.2 A photodiode is illuminated with a green beam (532 nm) whose power is unknown. The photodiode is operating in the photovoltaic regime at room temperature. After illumination, the voltage induced in the photodiode is 34 mV. Calculate the incident power if the quantum efficiency is 0.65 and if the electrical current generated in the photodiode in the absence of illumination is 1 mA.

3.3 Calculate the current induced in a photodiode with an intrinsic quantum efficiency of 0.90 when it is illuminated at room temperature with a 0.35 mW light beam whose wavelength is 1140 nm. The photodiode is working in the photoconductor regime, and in the absence of illumination no electrical current is generated by this photodiode. What happens if the photodiode is cooled down to 5 °C?

3.4 A light beam of 100 mW is to be measured with the AsGa photocathode of Figure 3.18. Calculate the number of photoelectrons generated if the incident wavelength is 650 nm (assume an absorbance of 100 %).

3.5 Calculate the dark current intensity at room temperature ($T = 300$ K) of a metallic photocathode with the following characteristics: area $= 3$ cm^2, $e\phi = 1.05$ eV. If the quantum efficiency of this photocathode is 0.75, calculate the minimum power that can be detected (the wavelength of the incident beam is 808 nm). Discuss the improvements if the photocathode is cooled down to 260 K.

3.6 Now calculate the minimum light power that can be measured with a photomultiplier using the photocathode of Exercise 3.5 and with 10 *dynodes*, each of which has a secondary emission coefficient of $\delta = 6$. Estimate these minimum powers if the photocathode is cooled down to 5 °C. Assume a bandpass width of 1 Hz.

3.7 The accelerating voltage in a streak camera is set to 500 V. If the distance between the electrodes used for electron deviations is 15 cm and the applied voltage between these electrodes is 1200 V, calculate the maximum spatial deviation that can be induced in the phosphor screen if the distance between the photocathode and the phosphor screen is 30 cm.

REFERENCES AND FURTHER READING

Guide for Spectroscopy, Jobin Yvon Spex Instruments S.A. Group.
Fernandez, J., Cussó, F., González, R., and García Solé, J. (eds.), Láseres sintonizables de estado sólido y aplicaciones, Ediciones de la Universidad Autónoma de Madrid (1989).
Kuzmany, H., *Solid State Spectroscopy. An Introduction*, Springer-Verlag, Berlin (1998).

4

The Optical Transparency of Solids

4.1 INTRODUCTION

In the previous chapters, we have examined the fundamentals of the most usual spectroscopic techniques, as well as the instrumentation needed in order to employ these techniques. We defined a number of optical magnitudes (such as α, A, T, and R), which are directly measurable as a function of the incident light wavelength. Thus, at this point we know how to measure the transparency of a solid over the whole optical range. Now, it is time to invoke some models in order to predict and interpret the transparency (absorption, reflection, or transmission spectra) of a given material. For this purpose, we will first relate these optical magnitudes (α, A, T, and R) to other classical magnitudes of the material, such as the relative dielectric constant, ε, or the refractive index, n. Then, we will use simple microscopic and macroscopic models in order to analyze the frequency (wavelength) dependence of these magnitudes. By the time we reach the end of this chapter, we will be able to predict some optical spectra and classify solid materials as metals, semiconductors, or insulators according to their optical behavior. Afterwards, we will see how the shapes of these spectra are intimately related to their electronic structure and how, inversely, the optical spectra provide very useful information on the particular electronic structure of a given solid material.

4.2 OPTICAL MAGNITUDES AND THE DIELECTRIC CONSTANT

Our first task is to connect measurable optical magnitudes with the dielectric constant, which describes the response of a given material to an applied electric field. This

An Introduction to the Optical Spectroscopy of Inorganic Solids J. García Solé, L. E. Bausá, and D. Jaque
© 2005 John Wiley & Sons, Ltd ISBNs: 0-470-86885-6 (HB); 0-470-86886-4 (PB)

electric field is created by the electromagnetic wave propagating into the solid. In the optical range, as defined in Section 1.2, the wavelength of the electromagnetic radiation (200 nm $\leq \lambda \leq$ 3000 nm) is much larger than the interatomic distances in solids. This fact enables us to consider a solid as a continuous medium for the purpose of describing the propagation of the electromagnetic radiation and then to use a classical description. Considering, for simplicity, isotropy in the medium, the spatial and temporal dependence for the electric field of an electromagnetic wave with an angular frequency ω, propagating along the z direction, can be written as

$$E = E_0 e^{i(Nkz - \omega t)} \tag{4.1}$$

where $k = 2\pi/\lambda$ is the module of the light wave vector in vacuum, $|E_0|$ is the electric field amplitude at $z = 0$, and N is the *complex refractive index*, defined by

$$N = n + i\kappa \tag{4.2}$$

The real part of this number is the *normal refractive index* $n = c/v$ (c and v being the speed of light in vacuum and in the medium, respectively). The imaginary part of the complex refractive index, κ, is called the *extinction coefficient*. It is necessary to recall here that both magnitudes, n and κ, are dependent on the frequency (wavelength) of the propagating wave ω, $N \equiv N(\omega)$.

Taking into account Equation (4.2) and that $k = \omega/c$, Equation (4.1) can be rewritten as

$$E = E_0 e^{-\frac{\omega}{c}\kappa z} e^{i\omega(\frac{n}{c}z - t)} \tag{4.3}$$

so that a nonzero extinction coefficient κ leads to an exponential attenuation of the wave in the material. As the intensity of the light wave, I, is proportional to the square of the electric field module ($I \sim |E|^2 = EE^*$), we can write an expression for the intensity attenuation along the propagating direction z:

$$I = I_0 e^{-2\frac{\omega}{c}\kappa z} \tag{4.4}$$

where I_0 is the intensity of the incident light ($z = 0$). Now, comparing this equation to the Lambert–Beer law (see Equation (1.4)), we obtain the relation between the absorption coefficient α, which is directly measurable from an absorption spectrum, and the extinction coefficient κ:

$$\alpha = \frac{2\omega}{c}\kappa \tag{4.5}$$

Let us now relate the refractive index and the extinction coefficient with the relative dielectric constant of the solid material. Assuming a nonmagnetic solid (relative magnetic permeability $\mu = 1$) we know that $N = \sqrt{\varepsilon}$, where ε is the relative dielectric constant of the material. Thus, as ε is also a complex magnitude, $\varepsilon = \varepsilon_1 + i\varepsilon_2$, we can write

$$(n + i\kappa)^2 = \varepsilon_1 + i\varepsilon_2 \tag{4.6}$$

and this equation can be solved for the real and imaginary terms as follows:

$$\varepsilon_1 = n^2 - \kappa^2 \tag{4.7}$$

$$\varepsilon_2 = 2n\kappa \tag{4.8}$$

It is known that measuring the absorption coefficient (and thus the extinction coefficient) over the whole frequency range, $0 \le \omega < \infty$, the real part of $N(\omega)$ – that is, the normal refractive index $n(\omega)$ – can be obtained by using the Kramers–Krönig relationships (Fox, 2001). This is an important fact, because it allows us to obtain the frequency dependence of the real and imaginary dielectric constants from an optical absorption experiment.

Inversely, we can obtain n and κ as functions of the relative dielectric constants:

$$n = \left[\tfrac{1}{2}\left((\varepsilon_1^2 + \varepsilon_2^2)^{1/2} + \varepsilon_1\right)\right]^{1/2} \tag{4.9}$$

$$\kappa = \left[\tfrac{1}{2}\left((\varepsilon_1^2 + \varepsilon_2^2)^{1/2} - \varepsilon_1\right)\right]^{1/2} \tag{4.10}$$

The reflectivity of a solid can also be determined after establishing the boundary conditions for the electromagnetic radiation at the interface between the solid and the vacuum. In the simple case of a solid in a vacuum, and considering normal incidence of light, it is well known from basic optics texts that

$$R = \frac{(1 - n)^2 + \kappa^2}{(1 + n)^2 + \kappa^2} \tag{4.11}$$

Thus, the optical magnitudes n, κ (or α), and R can be obtained from ε_1 and ε_2 by using Equations (4.9), (4.10), and (4.11), respectively.

In the next section, we will develop a simple model to predict the frequency dependence of the relative dielectric constants ε_1 and ε_2 of a given material. At that point, we will be able to determine the measurable optical magnitudes defined in Chapter 1 at any particular wavelength (or frequency) if the relative dielectric constants (and thus n and κ) are known at that wavelength.

EXAMPLE 4.1   *The complex refractive index of germanium at 400 nm is*
$N = 4.141 + i\,2.215$. At this wavelength, determine (a) the optical density for
a sample of thickness 1 mm and (b) the reflectivity at normal incidence.

(a) According to Equation (4.2), $\kappa = 2.215$ and, taking into account that $\omega = 2\pi c/\lambda$, the absorption coefficient is given by Equation (4.5):

$$\alpha = \frac{2\omega}{c}\kappa = \frac{4\pi}{\lambda}\kappa = \frac{4\pi \times 2.215}{400 \times 10^{-9}} = 6.96 \times 10^{7}\,\mathrm{m}^{-1} = 6.96 \times 10^{5}\,\mathrm{cm}^{-1}$$

and, according to Equation (1.10), the optical density for a 1 mm sample is

$$OD = \frac{\alpha x}{2.303} = \frac{6.96 \times 10^{5} \times 0.1}{2.303} = 30\,221$$

This is indeed a very large value to be measured by an optical spectrophotometer (see Section 1.3).

(b) From Equation (4.2), we know that $n = 4.141$ and so, using Equation (4.11), the reflectivity at normal incidence is

$$R = \frac{(1 - 4.141)^{2} + 2.215^{2}}{(1 + 4.141)^{2} + 2.215^{2}} = 0.471 = 47.1\,\%$$

As for the previous example, the optical transparency of a particular solid can be predicted from the magnitude of n and κ (or ε_1 and ε_2) at different wavelengths. These optical magnitudes are reported in different handbooks of optical materials (see, for instance, Palik, 1985; Weber, 2003). We will now mention some general aspects regarding the transparency of metals, semiconductors, and insulators, in the different spectral regions of the optical range (UV, 50–350 nm; visible, 350–700 nm; and near infrared, 700–3000 nm) by considering three relevant solids:

- Al (a typical metal) has a reflectivity R that varies from about 0 to 90 % in the UV, while it is highly reflecting in the visible and near infrared.

- Si (a typical semiconductor) is highly absorbing in the UV, and partially reflecting and absorbing in the visible and near infrared.

- SiO_2 (a typical insulator) has a strong absorption rise (called the fundamental absorption edge) in the UV and it is transparent in the visible.

4.3 THE LORENTZ OSCILLATOR

In the previous section we have demonstrated how the measurable optical magnitudes are related to the dielectric constants (ε_1, ε_2). Now we need to establish how these dielectric constants depend on the frequency of the incoming electromagnetic radiation. Then, we will be able to predict the particular shape of optical absorption and reflectivity spectra of any material.

Obviously, we need to start from microscopic (classic and quantum) models. These models require some knowledge about the nature of the interatomic (or interionic) bonding forces in our solid and whether or not the valence electrons are free to move inside the solid.

In metals, valence electrons are conduction electrons, so they are free to move along the solid. On the contrary, valence electrons in insulators are located around fixed sites; for instance, in an ionic solid they are bound to specific ions. Semiconductors can be regarded as an intermediate case between metals and insulators; valence electrons can be of both types, free or bound.

The most simple, but general, model to describe the interaction of optical radiation with solids is a classical model, due to Lorentz, in which it is assumed that the valence electrons are bound to specific atoms in the solid by harmonic forces. These harmonic forces are the Coulomb forces that tend to restore the valence electrons into specific orbits around the atomic nuclei. Therefore, the solid is considered as a collection of *atomic oscillators*, each one with its characteristic natural frequency. We presume that if we excite one of these atomic oscillators with its natural frequency (the *resonance frequency*), a resonant process will be produced. From the quantum viewpoint, these frequencies correspond to those needed to produce valence band to conduction band transitions. In the first approach we consider only a unique resonant frequency, ω_0; in other words, the solid consists of a collection of equivalent atomic oscillators. In this approach, ω_0 would correspond to the *gap frequency*.

This model of atomic oscillators, in which we assume bound valence electrons, is also perfectly valid for metals, except that in this case we must set $\omega_0 = 0$.

Let us now analyze the interaction of a light wave with our collection of oscillators at frequency ω_0. In this case, the general motion of a valence electron bound to a nucleus is a damped oscillator, which is forced by the oscillating electric field of the light wave. This atomic oscillator is called a *Lorentz oscillator*. The motion of such a valence electron is then described by the following differential equation:

$$m_e \frac{d^2\mathbf{r}}{dt^2} + m_e \Gamma \frac{d\mathbf{r}}{dt} + m_e \omega_0{}^2 \mathbf{r} = -e\mathbf{E}_{\text{loc}} \tag{4.12}$$

where m_e and e are the electronic mass and charge, respectively, and $\mathbf{r}$ is the electron's position with respect to equilibrium. The harmonic term $m_e \omega_0{}^2 \mathbf{r}$ represents a Hookes' law restoring force acting on the valence electron. Obviously, this term vanishes for metals as valence (conduction) electrons become free. The damping term in Equation (4.12), $m_e \Gamma dr/dt$, represents a viscous force due to the effect of the solid on the motion of the valence electrons, where Γ is the damping rate. This term arises from various scattering processes experienced by the valence electrons. In solids, it is typically related to loss of energy due to the excitation of phonons. The term $-e\mathbf{E}_{\text{loc}}$ is the force acting on the valence electron due to the oscillating electric field of the light $\mathbf{E}_{loc}$, where the subscript 'loc' indicates a local field. Assuming that this field oscillates as $e^{-i\omega t}$, the solution of Equation (4.12) is also time dependent and is given by

$$\mathbf{r} = \frac{-e\mathbf{E}_{loc}/m_e}{(\omega_0^2 - \omega^2) - i\Gamma\omega} \qquad (4.13)$$

where $\mathbf{r}$ is a complex number because of a phase shift between $\mathbf{r}$ and $\mathbf{E}_{loc}$ caused by the nonvanishing damping term.

Due to the oscillating nature of the local electric field, $\mathbf{E}_{loc}$, an oscillating electric dipole moment $\mathbf{p} = -e\mathbf{r}$ is induced. Taking into account that $\mathbf{p} = \alpha\mathbf{E}_{loc}$, α being the *atomic polarizability* (a relation that holds for not very high local electric fields), and using Equation (4.13), we obtain:

$$\alpha = \frac{e^2/m_e}{(\omega_0^2 - \omega^2) - i\Gamma\omega} \qquad (4.14)$$

Once again, it should be noted that the polarizability is a complex number because of the damping term.

At this point, we can relate the atomic polarizability α (a microscopic magnitude) with the dielectric constant ε (a macroscopic magnitude) if we make certain assumptions about the local electric field.

Remember that $\varepsilon = \mathbf{D}/\mathbf{E}$, where the displacement vector is given by $\mathbf{D} = \mathbf{E} + 4\pi\mathbf{P}$ (CGS units), $\mathbf{E}$ being the macroscopic electric field and $\mathbf{P}$ the macroscopic polarization. Considering a density of atoms N, the macroscopic polarization is $\mathbf{P} = N\langle\mathbf{p}\rangle = N\alpha\langle\mathbf{E}_{loc}\rangle$ (where the symbol $\langle\ \rangle$ indicates an average value) and so $\mathbf{D} = \mathbf{E} + 4\pi N\alpha\langle\mathbf{E}_{loc}\rangle$. Now assuming $\langle\mathbf{E}_{loc}\rangle = \mathbf{E}$, we obtain:[1]

$$\varepsilon = 1 + 4\pi N\alpha \qquad (4.15)$$

where it should be recalled that both α and ε are complex quantities. Inserting Equation (4.14) into Equation (4.15), we obtain the dependence of ε on the frequency of the incident light (CGS units):

$$\varepsilon = 1 + \frac{4\pi Ne^2}{m_e}\frac{1}{(\omega_0^2 - \omega^2) - i\Gamma\omega} \qquad (4.16)$$

and then, multiplying and dividing the second term on the right-hand side of this equation by $(\omega_0^2 - \omega^2) + i\Gamma\omega$, the real and imaginary dielectric constants are

---

[1] In general, $\langle\mathbf{E}_{loc}\rangle \neq \mathbf{E}$, since the local electric field is averaged over the atomic sites and not over the spaces between these sites. In metals, where valence electrons are free (nonlocalized electrons), the assumption $\langle\mathbf{E}_{loc}\rangle = \mathbf{E}$ is reasonable, but for bound valence electrons (dielectrics and semiconductors) this relation needs to be known. However, for our purpose of a qualitative description of optical properties, we will still retain this assumption.

obtained:

$$\varepsilon_1 = 1 + \frac{4\pi N e^2}{m_e} \frac{(\omega_0{}^2 - \omega^2)}{(\omega_0{}^2 - \omega^2)^2 + \Gamma^2 \omega^2} \tag{4.17}$$

$$\varepsilon_2 = \frac{4\pi N e^2}{m_e} \frac{\Gamma \omega}{(\omega_0{}^2 - \omega^2)^2 + \Gamma^2 \omega^2} \tag{4.18}$$

In the general case of more than one valence electron per atom, and allowing for the possibility of different resonance frequencies instead of a unique frequency ω_0, expression (4.16) is generalized to

$$\varepsilon = 1 + \frac{4\pi N e^2}{m_e} \sum_j \frac{N_j}{(\omega_j{}^2 - \omega^2) - i\Gamma_j \omega} \tag{4.19}$$

where N_j is the density of valence electrons bound with a resonance frequency ω_j and $\sum_j N_j = N$ is the total density of valence electrons.

The corresponding quantum mechanical expression of $\varepsilon(\omega)$ in Equation (4.19) is similar except for the quantity N_j, which is replaced by $N f_j$. However, the physical meaning of some terms are quite different: ω_j represents the frequency corresponding to a transition between two electronic states of the atom separated by an energy $\hbar \omega_j$, and f_j is a dimensionless quantity (called the *oscillator strength* and formally defined in the next chapter, in Section 5.3) related to the quantum probability for this transition, satisfying $\sum_j f_j = 1$. At this point, it is important to mention that the multiple resonant frequencies ω_j could be related to multiple valence band to conduction band singularities (transitions), or to transitions due to optical centers. This model does not differentiate between these possible processes; it only relates the multiple resonances to different resonance frequencies.

We are now able to understand the response of our solid to an electromagnetic field oscillating at frequency ω. For the sake of simplicity, we return to the use of expressions (4.17) and (4.18), related to a solid made of single-electron classical atoms, and to only one resonant frequency ω_0, related to the band gap. Using these expressions, in Figure 4.1(a) we have displayed the dependencies of ε_1 and ε_2 on the incident photon energy.

At the same time, Equations (4.9), (4.10), and (4.11) can now be used to obtain the spectral behavior of the measurable magnitudes n, κ, and R. Figure 4.1(b) shows the spectral dependencies of the extinction coefficient $\kappa(\omega)$ and the reflectivity $R(\omega)$, obtained from the values of ε_1 and ε_2 given in Figure 4.1(a). The spectral behavior of n does not have the same relevance as $\kappa(\omega)$ and $R(\omega)$ from the spectroscopic point of view, and has been omitted in Figure 4.1(b) for the sake of simplicity. Inspection of Figures 4.1(a) and 4.1(b) shows four significant spectral regions.

Spectral region I corresponds to illuminating frequencies that are much lower than the resonance frequency, $\omega \ll \omega_0$. According to Equations (4.18), and (4.8), in this region $\varepsilon_2 = 2n\kappa \approx 0$ and then (as $n \geq 1$) $\kappa \approx 0$. Also, in accordance with

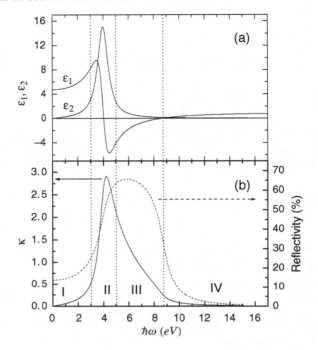

Figure 4.1 The spectral dependencies of (a) ε_1 and ε_2 and (b) κ and R. These curves have been represented for typical values within the optical range; $\hbar\omega_0 = 4$ eV, $\hbar\Gamma = 1$ eV, and $4\pi N e^2/m_e = 60$.

Equation (4.7), $\varepsilon_1 = n^2 - \kappa^2 \approx n^2 > 1$. We can see in Figure 4.1(b) that this spectral region is characterized by a high transparency; weak absorption and relatively low reflectivity. Figure 4.1(a) shows that in this region ε_1 increases with the frequency of the electromagnetic radiation, ω. The refractive index n (not displayed for the sake of simplicity), also increases with ω. This behavior corresponds to the so-called *normal dispersion*.

Spectral region II corresponds to frequencies close to the resonance frequency, $\omega \approx \omega_0$. Figure 4.1(a) shows that, in this region, ε_1 reaches a maximum, then decreases with the frequency of light in a certain energy range and increases again. The mentioned decrease is a manifestation of an *anomalous dispersion* in this spectral region. This anomalous behavior (also shown by n) is accompanied by an appreciable increase in the reflectivity $R(\omega)$ and a strong absorption peak in $\kappa(\omega)$ (and in $\varepsilon_2(\omega)$), as shown in Figure 4.1(b) (or Figure 4.1(a)).

In region III, $\omega \gg \omega_0$, the photon energy is much greater than the binding energy of the valence electron; the valence electrons respond as conduction electrons in metals (free electrons). Therefore, the insulators have a metallic reflectance (high reflectivity) in this spectral region. Figure 4.1(b) shows that spectral region III is consistent with a high reflectivity $R(\omega)$, while the extinction coefficient $\kappa(\omega)$ shows a rapid decrease with increasing frequency.

Finally, the onset of spectral region IV is defined by $\varepsilon_1 = 0$, and hence for frequencies $\omega > \omega_p$, where ω_p is called the *plasma frequency*. This frequency indicates a region of very high transparency ($\kappa \approx 0$ and $R \approx 0$), as shown in Figure 4.1(b), and is especially significant for metales, as we will see in the next section. In fact, at these very high frequencies the solid cannot respond to the electromagnetic field and becomes transparent. Letting $\varepsilon_1 = 0$ in Equation (4.17) and assuming $\omega \gg \omega_0 \gg \Gamma$, the plasma frequency is given (in CGS units) by

$$\omega_p = \sqrt{4\pi N e^2 / m_e} \qquad (4.20)$$

Let us now check the validity of the simple Lorentz model in order to explain the spectra of real solids. Figure 4.2 shows the dependence of the reflectivity on photon energy for a typical semiconductor, Si (Figure 4.2(a)), and for a typical insulator, KCl (Figure 4.2(b)). The Lorentz oscillator cannot quantitatively explain both spectra. In fact, we have supposed a single resonance frequency ω_0, but in the most general case a

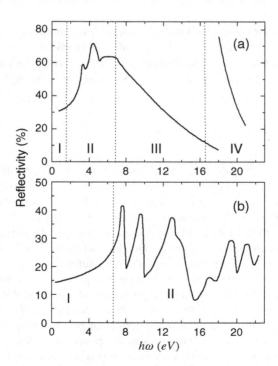

Figure 4.2   The reflectivity spectra of (a) Si and (b) KCl (reproduced with permission from Philipp and Ehrenreich, 1963a).

solid consists of a collection of oscillators with different resonance frequencies, spread over energy bands. Indeed, more than one resonant frequency is clearly observed in the spectra of both Si and KCl crystals.

In any case, the simple Lorentz model provides a *qualitatively* valid explanation of the main features of the spectra of solids, if we associate the resonance frequency ω_0 with the transition frequency corresponding the main band-gap transition of the solid. In this respect, the reflectivity spectrum of Si (Figure 4.2 (a)) is easily explained by the Lorentz model. The four spectral regions are clearly observed, although only region I and, partially, region II lie in the optical range. The sharp rise at about 3 eV leads to a peak frequency of $\omega_0 \simeq 6.8 \times 10^{15} \text{s}^{-1}$. Considering four valence electrons per Si atom, Equation (4.17) leads to a value of $\varepsilon_1 = 13.5$ for low frequencies ($\omega \to 0$). Indeed, this value is not far from that experimentally measured at low frequencies, $\varepsilon_1 = 12$.

Regarding KCl, spectral regions I and II are clearly visible (see Figure 4.2(b)). As region I covers all of the optical range, potassium chloride is a highly transparent material from the spectroscopic viewpoint. At energies higher than about 7 eV, a number of sharp peaks occur. These peaks correspond to excitons and band-to-band singularities, and they cannot be quantitatively explained by the classical Lorentz oscillator. However, the qualitative response is satisfactorily explained, as spectral regions I and II are observed.

In general, spectral regions III and IV are far away from the optical range for good insulators, while these regions can be optically observed for some semiconductors. This is the reason why many semiconductors, such as Ge and Si, have a metallic aspect, while most of the good insulators, such as KCl and NaCl, are highly transparent in the visible.

4.4 METALS

We will now analyze the general optical behavior of a metal using the simple Lorentz model developed in the previous section. Assuming that the restoring force on the valence electrons is equal to zero, these electrons become free and we can consider that $\omega_0 = 0$ in Equation (4.12). This is the so-called *Drude model*, which was proposed by P. Drude in 1900. We will see how this model successfully explains a number of important optical properties, such as the fact that metals are excellent reflectors in the visible while they become transparent in the ultraviolet.

The previous assumption, $\omega_0 = 0$, can be also justified from the quantum point of view, as the most relevant transitions in metals take place within a band, usually the conduction band. The energy levels within a band of a solid are separated by energies of about 10^{-27} eV, and so in the optical range (photon energies of some eV) the validity of assuming $\omega_j \cong 0$ for transitions between these levels is evident. Consequently, the classical assumption $\omega_0 = \omega_j \cong 0$ is also justified from the quantum viewpoint.

4.4.1 Ideal Metal

To simplify, we begin thinking of an *ideal metal* – that is, a metal without damping – and later we shall discuss how damping forces can affect the optical properties of this ideal metal. Thus, setting $\omega_0 = 0$ and $\Gamma = 0$ in Equations (4.17) and (4.18), we obtain

$$\varepsilon_1 = 1 - \frac{4\pi N e^2}{m_e} \frac{1}{\omega^2} = 1 - \frac{\omega_p^2}{\omega^2} \tag{4.21}$$

$$\varepsilon_2 = 0 \tag{4.22}$$

where ω_p is the plasma frequency given in Equation (4.20).

Using expressions (4.9) and (4.10), the optical magnitudes n and κ can be obtained as functions of the frequency of the illuminating light. Figure 4.3 shows a plot of the optical magnitudes as functions of ω/ω_p. For light frequencies lower than ω_p, the refractive index is zero ($n = 0$), while the extinction coefficient decreases with frequency [$\kappa^2 = (\omega_p^2/\omega^2) - 1$], being $\kappa = 0$ for $\omega = \omega_p$. For frequencies higher than the plasma frequency, the extinction coefficient remains equal to zero ($\kappa = 0$), while the refractive index rises with increasing frequency [$n^2 = 1 - (\omega_p^2/\omega^2)$] toward a limit $n = 1$. Obviously, these values of the refractive index are not physically acceptable. For instance, a zero value for n leads to an infinite phase velocity ($v = c/n$). The physical meaning of this infinite phase velocity is that all valence electrons are oscillating in phase for frequencies below ω_p, while for frequencies larger than ω_p this coherence is broken and the plasma is formed.

Figure 4.4 shows the optical spectra expected for our ideal metal, calculated from expressions (4.5) and (4.11). The absorption spectrum shows a rapid decrease in the

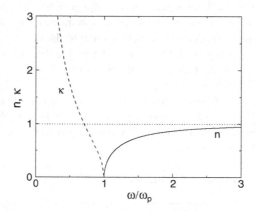

Figure 4.3 The optical magnitudes, n and κ, of an ideal metal (undamped free electrons) versus the light frequency.

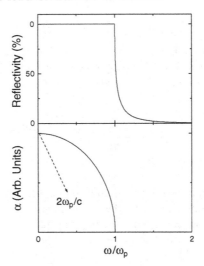

Figure 4.4   Reflectivity and optical absorption spectra for an ideal metal.

absorption coefficient, α, with increasing frequency [$\alpha = (2\omega/c)\sqrt{(\omega_p^2/\omega^2) - 1}$] and becomes zero for $\omega \geq \omega_p$. The reflectivity spectrum shows that $R = 1$ for frequencies lower than ω_p and decreases rapidly ($R = [\omega_p/(\omega + \sqrt{\omega^2 - \omega_p^2})]^4$) for frequencies higher than ω_p.

Thus, the Drude model predicts that ideal metals are 100 % reflectors for frequencies up to ω_p and highly transparent for higher frequencies. This result is in rather good agreement with the experimental spectra observed for several metals. In fact, the plasma frequency ω_p defines the region of transparency of a metal. It is important to realize that, according to Equation (4.20), this frequency only depends on the density of the conduction electrons N, which is equal to the density of the metal atoms multiplied by their valency. This allows us to determine the region of transparency of a metal provided that N is known, as in the next example.

EXAMPLE 4.2   *Sodium is a metal with a density of conduction electrons* $N = 2.65 \times 10^{22}$ *cm^{-3}. Determine (a) its plasma frequency, (b) the wavelength region of transparency, and (c) the optical density at very low frequencies for a Na sample of 1 mm thickness.*

(a) According to Equation (4.20),

$$\omega_p = \sqrt{\frac{4\pi \times 2.65 \times 10^{22}\ \text{cm}^{-3} \times (1.6 \times 10^{-19} \times 3 \times 10^9\ \text{stc})^2}{9.1 \times 10^{-28}\ \text{g}}}$$

$$= 9.18 \times 10^{15}\text{s}^{-1}$$

(b) Then metallic sodium will become transparent for frequencies higher than 9.18×10^{15} s^{-1}. The corresponding wavelength λ_p, called the *cutoff wavelength*, is

$$\lambda_p = \frac{2\pi c}{\omega_p} = \frac{2\pi \times 3 \times 10^{10} \text{ cm s}^{-1}}{9.18 \times 10^{15} \text{ s}^{-1}} = 2.05 \times 10^{-5} \text{ cm} = 205 \text{ nm}$$

Consequently, Na is a good *optical filter* for wavelengths shorter than 205 nm.

(c) At low frequencies (long wavelengths), Na is a fully reflector material. The optical absorption coefficient is also very large. It is given by $\alpha = (2\omega/c)\sqrt{\omega_p^2/\omega^2 - 1}$. Thus, at $\omega = 0$:

$$\alpha = \frac{2\omega_p}{c} = \frac{2 \times 9.18 \times 10^{15} \text{ s}^{-1}}{3 \times 10^{10} \text{ cm s}^{-1}} = 6.12 \times 10^5 \text{ cm}^{-1}$$

and according to Equation (1.10), the corresponding optical density for a sample of thickness 1 mm is

$$OD = 6.12 \times 10^5 \text{ cm}^{-1} \times 0.1 \text{ cm} \times \log e = 2.6 \times 10^4$$

Indeed, this optical density is too large to be measured with a spectrophotometer. Moreover, the ideal metal reflectivity is 1, so that no light would be transmitted at low frequencies.

In the previous example, we have calculated the plasma frequency for metallic Na from the free electron density N. In Table 4.1, the measured cutoff wavelengths, λ_p, for different alkali metals are listed together with their free electron densities. The relatively good agreement between the experimental values of λ_p and those calculated from Equation (4.20), within the ideal metal model, should be noted. It can also be observed that the N values range from about 10^{22} to about 10^{23} cm^{-3}, leading to

Table 4.1 Calculated and measured cutoff wavelengths for alkali metals

Metal	N ($\times 10^{22}$ cm^{-3})	λ_p, (nm), calculated	λ_p, (nm), experimental
Li	4.70	154	205
Na	2.65	205	210
K	1.40	282	315
Rb	1.15	312	360
Cs	0.91	350	440

calculated cutoff wavelengths inside the UV spectral region. Thus, in general, metals are good *filters* for the UV radiation, transmitting radiation with wavelengths shorter than λ_p while reflecting (and absorbing too) radiation with wavelengths larger than λ_p. In some ways, this property of metals is similar to the high reflectivity of the ionosphere (due to the high concentration of free electrons) for frequencies up to about 3 MHz (radio waves), which makes signal transmission over long distances possible.

4.4.2 Damping effects

If we now consider a nonvanishing damping term, $\Gamma \neq 0$, Equations (4.21) and (4.22) must be written as:

$$\varepsilon_1 = 1 - \frac{\omega_p^2}{\omega^2 + \Gamma^2} \tag{4.23}$$

$$\varepsilon_2 = \frac{\omega_p^2 \Gamma}{\omega(\omega^2 + \Gamma^2)} \tag{4.24}$$

where we can see that, at variance with the ideal metal, ε_2 differs from zero.

The damping term in a metal is due to the scattering suffered by the free electrons with atoms and electrons in the solid, which produces the electrical resistivity.

Let us now imagine the motion of the free electrons just after the driving external local field is eliminated. Then Equation (4.12) for the Lorentz oscillator appears in a simplified form, as

$$\frac{d^2\mathbf{r}}{dt^2} + \Gamma \frac{d\mathbf{r}}{dt} = 0 \tag{4.25}$$

which can be written in terms of the free electron velocity, v, as

$$\frac{dv}{dt} + \Gamma v = 0 \tag{4.26}$$

The solution of this differential equation is $v = v_0 e^{-\Gamma t}$. This solution indicates that the electron velocity decays exponentially to zero with a decay time of $\tau = 1/\Gamma$ after the driving force (electric local field) is eliminated. This time represents the *mean free collision time* for electrons in metals, which is typically $\tau \approx 10^{-14}$ s, and corresponds to a damping frequency of $\Gamma \approx 10^{14}$ s^{-1}. This means that damping effects will be significant for frequencies $\leq 10^{14}$ s^{-1}. In other words, these effects could be important for wavelengths larger than about $\lambda = c/\Gamma = 3000$ nm; that is, at the long wavelength limit of the optical range. Therefore, we expect that our ideal model metal would not be affected by damping effects as far as optical spectroscopy is concerned.

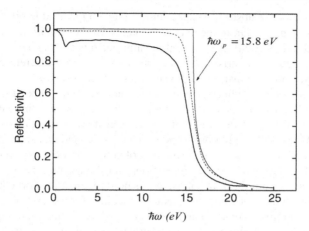

Figure 4.5 The reflectivity spectrum of Aluminum (full line) compared with those predicted from the ideal metal model with $\hbar\omega_p = 15.8$ eV (dotted line) and a damped oscillator with $\Gamma = 1.25 \times 10^{14}$ s^{-1} (dashed line) (experimental data reproduced with permission from Ehrenreich *et al.*, 1962).

In Figure 4.5, the experimental reflectivity spectrum of aluminum is compared with those predicted by the ideal metal and the damped metal models. Al has a free electron density of $N = 18.1 \times 10^{22}$ cm^{-3} (three valence electrons per atom) and so, according to Equation (4.20), its plasma energy is $\hbar\omega_p = 15.8$ eV. Thus, the reflectivity spectrum for the ideal metal can be now calculated. Compared to the experimental spectrum, the ideal metal model spectrum is only slightly improved when taking into account the damping term, with $\Gamma = 1.25 \times 10^{14}$ s^{-1}, a value deduced from DC conductivity measurements. The main differences between the two calculated spectra are that damping produces a reflectivity slightly less than one below ω_p and the ultraviolet transmission edge is slightly smoothed out.

Finally, it should be mentioned that neither the ideal metal model nor the damped metal model are able to explain why the actual reflectivity of aluminum is lower than the calculated one ($R \approx 1$) at frequencies lower than ω_p. Also, these simple models do not reproduce features such as the reflectivity dip observed around 1.5 eV. In order to account for these aspects, and then to have a better understanding of real metals, the band structure must be taken into account. This will be discussed at the end of this chapter, in Section 4.8.

4.5 SEMICONDUCTORS AND INSULATORS

Unlike metals, semiconductors and insulators have bound valence electrons. This aspect gives rise to *interband transitions*. The objective of this and the next section is to understand how the absorption spectrum of a given material is related to its band structure and, in particular, to the density of states for the transitions involved.

First of all, let us recall the physical meaning of *band structure* in a solid. Transitions in isolated atoms arise from a series of states with discrete energy levels. Consequently, optical transitions between these levels give rise to absorption and emission lines at specific photon resonance frequencies. These spectra correspond to the case of individual atoms, or of widely separated atoms (at infinite distances). As the interatomic distance is decreased, the individual atomic charge distributions begin to interact. As a result, each atomic energy level shifts and can split into $(2l + 1)N$ molecular energy levels, N being the number of atoms (or ions) involved in the bonding and $(2l + 1)$ being the orbital degenerancy of the atomic energy level (l being the orbital quantum number). In solids, atoms hold together at short equilibrium distances mostly because of ionic, covalent, or metallic forces, and N is of the order of Avogadro's number. Consequently, each energy level splits into a high number of closely spaced levels, giving rise to a *continuous band*. In any case, in spite of the fact that optical transitions in solids are much more complicated than atomic transitions, some features still retain the character of the transitions for individual atoms.

Figure 4.6 provides a good qualitative scheme to explain the metallic or insulator character of a solid.

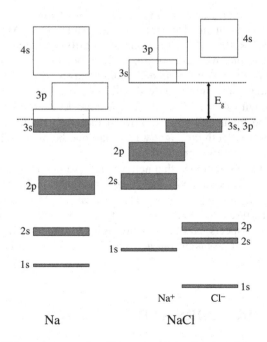

Figure 4.6 The filling of energy bands in a metal (sodium) and in an insulator (sodium chloride).

For atomic (gas) sodium (Na), the electronic configuration is $1s^2 2s^2 2p^6 3s^1$, leading to filled electronic energy levels 1s, 2s and 2p, while the 3s level is half-filled. The other excited levels, 3p, 4s..., are empty. In the solid state (the left-hand side in Figure 4.6), these atomic energy levels are shifted and split into energy bands: bands 1s, 2s and 2p are fully occupied, while the 3s ($l = 0$) band, the conduction band, is half-filled, so that a large number ($N(2l + 1)/2 = N/2$) of empty 3s excited levels is still available. As a result, electrons are easily excited into empty levels by an applied electric field, and so become free electrons. This aspect confers the typical metallic character to solid sodium.

The case of sodium chloride (NaCl) (the right-hand side in Figure 4.6), a typical insulator, is quite different. The electronic configuration of Cl atoms is $1s^2 2s^2 2p^6 3s^2 3p^5$. However, sodium chloride is an ionic crystal and its crystal structure is constituted by Na^+ and Cl^- ions. This is because, at the short equilibrium distances in the NaCl solid, each Na atom prefers to transfer its 3s electron (see its electronic configuration above) to a 3p level of a Cl atom. This process leaves the 3s band of Na^+ ions empty, whereas the 3p band of Cl^- ions is fully occupied (notice that this band is mixed with the 3s Cl^- band). The energy difference between the top of the 3s 3p (Cl^-) band (the valence band) and the bottom of the 3s (Na^+) band (the conduction band) is about 8 eV and is known as the *band-gap energy*. Consequently, electrical conductivity by electronic transport does not occur in NaCl when an electric field is applied. From the optical viewpoint, the appearance of this energy gap gives rise to a continuous absorption spectrum for energies higher than about 8 eV for the NaCl crystal. This spectrum is due to *interband transitions*, from the 3s 3p (Cl^-) states to excited states of the 3s band (Na^+) and states of other higher energy bands. For energies lower than about 8 eV, sodium chloride does not absorb, so that this material is transparent in the visible and in the ultraviolet up to a *wavelength gap* of about $\lambda_g = 155$ nm (in the UV vacuum spectral region).

Thus, the energy gap gives the region of transparency for a nonmetallic solid. Table 4.2 gives the energy-gap E_g values (and the corresponding wavelengths, λ_g) for typical semiconductors and insulators. The low energy-gap values ($E_g \sim 1$ eV) for Ge, Si, and AsGa, explain the semiconductor character observed at room temperature for these solids, as sufficient valence electrons can be thermally excited across the band gap. These low energy-gap values also explain their opaque aspect in the visible

Table 4.2   The energy gap (E_g) and wavelength gap (λ_g) for a variety of solids

Solid	E_g (eV)	λ_g (nm)
Ge	0.67	1851
Si	1.14	1088
GaAs	1.5	827
ZnO	3.2	387
Diamond	5.33	233
LiF	13.7	90

spectral region. On the other hand, the large energy-gap values of ZnO, diamond, and LiF make these crystals highly transparent in the visible. In particular, the high energy gap of LiF makes this crystal of particular interest for UV-visible windows.

The simple energy-gap scheme of Figure 4.6 seems to indicate that transitions in solids should be broader than in atoms, but still centered on defined energies. However, interband transitions usually display a complicated spectral shape. This is due to the typical band structure of solids, because of the dependence of the band energy E on the wave vector $\mathbf{k}$ ($|\mathbf{k}| = 2\pi/a$, a being an interatomic distance) of electrons in the crystal.

Figure 4.7(a) shows the band structure of silicon around the energy gap, $E_g = 1.14$ eV. The region that separates the valence band (VB) from the conduction band (CB) corresponds to the gap of this material. This band structure is responsible for several features in the valence to conduction band spectra. For instance, Figure 4.7(b) shows the room temperature absorption spectrum of silicon, which is dominated by two main peaks at photon energies of about $E_1 = 3.5$ eV and $E_2 = 4.3$ eV (compare this with the reflectivity spectrum given in Figure 4.2(a)) and has a very weak rise (not apparent in the figure) at about 1.14 eV. This latter feature is related to the energy gap, the onset of band-to-band transitions.

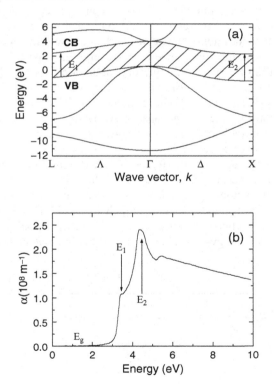

Figure 4.7 (a) The band structure diagram of Si near the gap energy (reproduced with permission from Cohen and Chelikowsky, 1988). (b) The room temperature absorption of Si (reproduced with permission from Palik, 1985).

The two peaks at E_1 and E_2 are related to transitions (indicated by arrows in Figure 4.7) from the vicinity of specific points of high symmetry in the Brillouin zone, denoted by L and X. However, in order to explain the exact locations of these two absorption peaks, the density of states function must be also taken into account. Thus, it appears obvious that the full interpretation of interband spectra is far from simple.

The absorption spectrum involving the valence band to first empty (conduction) band transitions is usually called the *fundamental absorption spectrum*. For many crystals, this spectrum lies within the optical range.

4.6 THE SPECTRAL SHAPE OF THE FUNDAMENTAL ABSORPTION EDGE

In the absorption spectrum of silicon (Figure 4.7(b)), we can appreciate that the absorption coefficient around the main peaks at 3.5 and 4.3 eV is extremely large, of the order of 10^8 m^{-1}. This implies very high optical densities (4.3×10^3 for a 0.1 mm thick sample), so that the absorption spectra can only be measured using thin films deposited over transparent substrates. In fact, very often it is only possible to examine the spectral region in the vicinity of the energy gap (around 1.14 eV for Si). This region is usually called the *fundamental absorption edge* and it is shown by a rapid rise in the absorption coefficient. The fundamental absorption edge provides very useful information on the band structure around the energy gap. In this section, we will examine the spectral shape of the fundamental absorption edge and its relationship with the band structure.

If a solid is illuminated with radiation of frequency $\omega \geq \omega_g$ ($E_g = \hbar\omega_g$), the radiation will become partially absorbed. The absorption coefficient at this frequency is proportional to:

- the probability P_{if} of a transition from the initial state i to the final state f (derived from the atomic orbitals of the constituent atoms);

- the density of electrons in the initial state n_i and the density of available final states n_f, including all of the states separated by an energy equal to $\hbar\omega$.

Therefore we can write:

$$\alpha(\omega) = A \sum_{i,f} P_{if} \, n_i n_f \qquad (4.27)$$

where A is a factor to allow agreement with the absorption coefficient units.

Before we deal with the spectral shape of $\alpha(\omega)$, let us examine the selection rule imposed by the conservation of momentum. For a given transition,

$$\mathbf{k}_p + \mathbf{k}_i = \mathbf{k}_f \qquad (4.28)$$

where $\mathbf{k}_i$ and $\mathbf{k}_f$ are the wave vectors of initial and final electron states, and $\mathbf{k}_p$ is the wave vector of the incoming photon. Since the magnitude of the wave vector for a typical visible photon ($\lambda = 600$ nm) is $k_p = 2\pi/\lambda \approx 10^5$ cm^{-1}, while for a typical electron in a crystal, k is related to the size of the Brillouin zone, $k_{i,f} \approx \pi/a \approx 10^8$ cm^{-1} (where we have considered a unit cell dimension of $a \cong 1$ Å $= 10^{-8}$ cm), we find that $k_p \ll k_{i,f}$, so that the selection rule (4.28) becomes

$$\mathbf{k}_i = \mathbf{k}_f \tag{4.29}$$

This important selection rule indicates that interband transitions must preserve the wave vector. Transitions that preserve the wave vector (such as those marked by vertical arrows in Figure 4.8(a)) are called *direct transitions*, and they are easily observed in materials where the top point in the valence band has the same wave vector as the bottom point in the conduction band. These materials are called *direct-gap materials*.

In materials with a band structure such as that sketched in Figure 4.8(b), the bottom point in the conduction band has a quite different wave vector from that of the top point in the valence band. These are called *indirect-gap* materials. Transitions at the gap photon energy are not allowed by the rule given in Equation (4.29), but they are still possible with the participation of lattice phonons. These transitions are called *indirect transitions*. The momentum conservation rule for indirect transitions can be written as

$$\mathbf{k}_i \pm \mathbf{k}_\Omega = \mathbf{k}_f \tag{4.30}$$

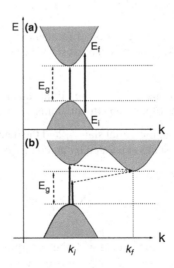

Figure 4.8 Interband transitions in solids with band-gap energy E_g: (a) A direct band gap. Two direct transitions are indicated by arrows. (b) An indirect band gap. Two indirect 'band-gap' transitions are indicated by arrows. The transitions at photon energies lower than E_g require absorption of phonons. The transitions at photon energies higher than E_g involve emission of phonons.

where $\mathbf{k}_\Omega$ denotes the wave vector of the phonons involved. The $\pm$ sign in Equation (4.30) indicates that indirect transitions can occur by absorption $(+)$ or by emission $(-)$ of phonons. In the first case, the crystal is illuminated with photons of energy $E = E_g - E_\Omega$, while in the second case the crystal is illuminated with photons of energy $E = E_g + E_\Omega$ (E_Ω being the energy of the phonon involved).

Indirect transitions are much weaker than direct transitions, because the latter do not require the participation of photons. However, many indirect-gap materials play an important role in technological applications, as is the case of silicon (band structure diagram in Figure 4.7(a)) or germanium (band structure diagram shown later, in Figure 4.11). Hereafter, we will deal with the spectral shape expected for both direct and indirect transitions.

4.6.1 The Absorption Edge for Direct Transitions

We will first consider transitions between two direct valleys, as in Figure 4.8(a). We suppose that all of the momentum-conserving transitions are allowed (*allowed direct transitions*). This means that P_{if} differs from zero for any value of $\vec{k}$. Assuming that P_{if}, which is related to the matrix elements of the initial and final electronic states, is independent of the frequency that connects these states (i.e., of $|\vec{k}|$), expression (4.27) can be rewritten as

$$\alpha(\omega) = A P_{if} \rho(\omega) \tag{4.31}$$

where $\rho(\omega)$ is the *joint density of states* (the number of pairs of states with the same wave vector $\mathbf{k}$, separated by $\hbar\omega$). This function depends on the shape of the energy bands.

In Appendix A1 we have determined that for the general case of parabolic bands, $\rho(\omega)$ is given by

$$\rho(\omega) = \frac{1}{2\pi^2} \left(\frac{2\mu}{\hbar} \right)^{3/2} (\omega - \omega_g)^{1/2} \tag{4.32}$$

where μ is the reduced effective mass, given by $1/\mu = 1/m_e^* + 1/m_h^*$ (m_e^* and m_h^* being the electron and hole effective masses, respectively). Thus, we expect the following frequency dependence for a *direct allowed absorption edge*:

$$\alpha(\omega) = 0, \qquad \text{for} \quad \omega < \omega_g$$
$$\alpha(\omega) \propto (\omega - \omega_g)^{1/2}, \qquad \text{for} \quad \omega \geq \omega_g \tag{4.33}$$

Several III–V semiconductors, such as AlP, GaAs, InSb, AlAs, and InAs, show direct absorption edge transitions. The next example shows the analysis of the fundamental absorption edge for indium arsenide.

EXAMPLE 4.3   *Figure 4.9(a) shows the dependence of the absorption coefficient versus the photon energy for indium arsenide. (a) Determine whether or not InAs is a direct-gap semiconductor. (b) Estimate the band-gap energy. (c) If an InAs sample of 1 mm thickness is illuminated by a laser of 1 W at a wavelength of 2 μm, determine the laser power for the beam after it passes through the sample. Only consider the loss of light by optical absorption.*

(a) To determine whether or not InAs is a direct-gap material, we calculate the square of the absorption coefficient α^2 versus the photon energy $\hbar\omega$ from the absorption spectrum given in Figure 4.9(a) This plot is shown in Figure 4.9(b); a linear dependence $\alpha^2 \propto (\hbar\omega - \hbar\omega_g)$ is clearly observed, in agreement with Equation (4.33). Consequently, we can say that InAs is a direct-gap semiconductor.

(b) The band-gap energy of InAs can be directly obtained by extrapolating the linear dependence in Figure 4.9(b) to zero, which leads to $E_g = \hbar\omega_g = 0.35$ eV.

(c) According to Figure 4.9(a), the absorption coefficient at 2 μm (0.62 eV) is 10^6 m^{-1}. Thus, we can obtain the attenuation of the laser intensity by using

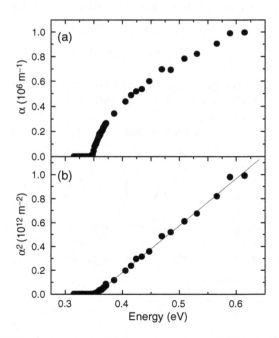

Figure 4.9   (a) The room temperature absorption spectrum of InAs in the fundamental absorption edge region. (b) The square of the absorption coefficient plotted versus the photon energy.

Equation (1.4). Assuming that the beam size is conserved, the attenuation of the laser power is given by

$$P = P_0 e^{-\alpha x} = (1 \text{ W}) \times e^{-10^6 \text{m}^{-1} \times 10^{-3} \text{m}} = e^{-1000} \text{ W} \approx 0 \text{ W}$$

so that the power is completely attenuated by absorption.

Reflectivity has not been considered here, but it should also be high, since we are in spectral region II of Figure 4.1(b). In an actual situation, only a laser power of $P_0(1 - R)$ would penetrate into the sample, R being the reflectivity of the sample. Taking into account the reflectivity in the inner face of the sample, the laser power after the beam passes through the sample would be $P = P_0 e^{-\alpha x}(1 - R)^2 = e^{-1000}(1 - R)^2$ W ≈ 0 W. In any case, the laser power is completely attenuated by the InAs sample.

For some direct-gap materials, the quantum electronic selection rules lead to $P_{if} = 0$. However, this is only strictly true at $k = 0$. For $k \neq 0$, it can be assumed, in a first order approximation, that the matrix element involving the top valence and the bottom conduction states is proportional to k; that is, $P_{if} \sim k^2$. Within the simplified model of parabolic bands (see Appendix A1), it is obtained that $\hbar\omega = \hbar\omega_g + \hbar^2 k^2 / 2\mu$, and therefore $P_{if} \sim k^2 \sim (\omega - \omega_g)$. Thus, according to Equations (4.31) and (4.32), the absorption coefficient for these transitions (called *forbidden direct transitions*) has the following spectral dependence:

$$\alpha(\omega) = 0, \quad \text{for} \quad \omega < \omega_g$$
$$\alpha(\omega) \propto (\omega - \omega_g)^{3/2}, \quad \text{for} \quad \omega \geq \omega_g \qquad (4.34)$$

Some complex oxides, such as SiO_2 or CuO_2, show forbidden direct transitions.

4.6.2 The Absorption Edge for Indirect Transitions

For indirect-gap materials, all of the occupied states in the valence band can be connected to all the empty states in the conduction band. In this case, the absorption coefficient is proportional to the product of the densities of initial states and final states (see Equation (4.27)), but integrated over all the possible combinations of states separated by $\hbar\omega \pm E_\Omega$ (E_Ω being the energy of the phonon involved). This formal calculation is beyond the scope of this book and leads to the following spectral dependence (Yu and Cardona, 1999) in the vicinity of ω_g:

$$\alpha(\omega) \propto (\omega - \omega_g \pm \Omega)^2 \qquad (4.35)$$

Table 4.3 The frequency dependence expected for the fundamental absorption edge of direct- and indirect-gap materials

Material	Frequency dependence
Direct allowed gap	$\alpha(\omega) \propto (\omega - \omega_g)^{1/2},$ for $\omega \geq \omega_g$
Direct forbidden gap	$\alpha(\omega) \propto (\omega - \omega_g)^{3/2},$ for $\omega \geq \omega_g$
Indirect gap	$\alpha(\omega) \propto (\omega - \omega_g \pm \Omega)^2$

where the term $\pm \Omega$ indicates whether a phonon of frequency Ω is absorbed or emitted.

It should be noted that the frequency dependence is different to those expected for direct-gap materials, given by Equations (4.33) and (4.34). This provides a convenient way of determining the direct or indirect nature of a band gap in a particular material by simply analyzing the fundamental absorption edge. Table 4.3 summarizes the frequency dependence expected for the fundamental absorption edge of direct- and indirect-gap materials.

The general shape of the absorption edge for an indirect-gap material has been sketched in Figure 4.10(a). In this figure (a plot of $\alpha^{1/2}$ versus ω), two different linear regimes are clearly observable. The straight line at lower frequencies shows an absorption threshold at a frequency of $\omega_1 = \omega_g - \Omega$, which corresponds to a process

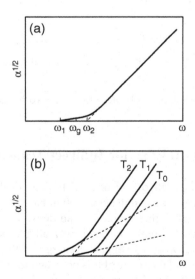

Figure 4.10 (a) The spectral shape expected for an indirect gap. (b) The temperature evolution expected for an indirect gap. $T_0 = 0$ K and $T_2 > T_1$.

involving absorption of phonons of energy $\hbar\Omega$. The second straight line intersects the frequency axis at $\omega_2 = \omega_g + \Omega$ and corresponds to a process in which phonons of energy $\hbar\Omega$ are emitted. The frequency gap corresponds to the midpoint between these two frequencies, $\omega_g = (\omega_1 + \omega_2)/2$. In addition, the frequency of the phonon involved can also be determined from the spectrum: $\Omega = (\omega_2 - \omega_1)/2$.

Because of the involvement of phonons in indirect transitions, one expects that the absorption spectrum of indirect-gap materials must be substantially influenced by temperature changes. In fact, the absorption coefficient must be also proportional to the probability of photon–phonon interactions. This probability is a function of the number of phonons present, η_B, which is given by the Bose–Einstein statistics:

$$\eta_B = \frac{1}{e^{\hbar\Omega/kT} - 1} \tag{4.36}$$

Thus, we have to introduce new proportionality factors in expression (4.35) to take this effect into account:

(i) For a transition with phonon absorption, $\alpha \propto \eta_B$.

(ii) For a transition with phonon emission, $\alpha \propto (\eta_B + 1)$.

Then, we can rewrite expression (4.35) in a more general way, which takes into account the temperature dependence of η_B:

$$\alpha_a \propto (\omega - \omega_g + \Omega)^2 \times \eta_B = \frac{(\omega - \omega_g + \Omega)^2}{e^{\hbar\Omega/kT} - 1} \tag{4.37}$$

$$\alpha_e \propto (\omega - \omega_g - \Omega)^2 \times (\eta_B + 1) = \frac{(\omega - \omega_g - \Omega)^2}{1 - e^{-\hbar\Omega/kT}} \tag{4.38}$$

where the subscripts a and e account for phonon absorption or phonon emission processes, respectively. Since both processes are possible, the absorption coefficient is given by $\alpha(\omega) = \alpha_a(\omega) + \alpha_e(\omega)$.

Figure 4.10(b) shows the temperature dependence of the absorption spectrum expected for an indirect gap. It can be noted that the contribution due to α_a becomes less important with decreasing temperature. This is due to the temperature dependence of the phonon density factor (see Equation (4.37)). Indeed, at 0 K there are no phonons to be absorbed and only one straight line, related to a phonon emission process, is observed. From Figure 4.10(b) we can also infer that ω_g shifts to higher values as the temperature decreases, which reflects the temperature dependence of the energy gap.

EXAMPLE 4.4   *The absorption edge of Ge.*

Germanium is an indirect band-gap material, but its band structure allows us to observe two energy gaps (one direct and one indirect) on analyzing the optical spectral region around the fundamental edge. As shown in Figure 4.11(a), the bottom location in the conduction band occurs at a point denoted by L_1, while the top point in the valence band (called the Γ_{25} point) is at $k = 0$. This leads to an indirect energy gap at 0.66 eV.

In Figure 4.11(b), a plot of $\alpha^{1/2}$ versus the photon energy, $\hbar\omega$, has been displayed from the absorption spectrum of Ge at 300 K. The double linear fit confirms the indirect nature of the $\Gamma_{25} \rightarrow L_1$ transition, as expected from Equations (4.37) and (4.38). In agreement with Figure 4.10(a), we extrapolate to find $\hbar\omega_1 = 0.618$ eV and $\hbar\omega_2 = 0.637$ eV. Now, we are able to estimate the band gap at room temperature, $\hbar\omega_g = (\hbar\omega_1 + \hbar\omega_2)/2 = 0.628$ eV, (not far from the value given by the band-gap energy, 0.66 eV, at 0 K) and the energy of the phonon assisting the indirect transition, $\hbar\Omega = (\hbar\omega_2 - \hbar\omega_1)/2 = 9.5 \times 10^{-3}$ eV ≈ 77 cm^{-1}.

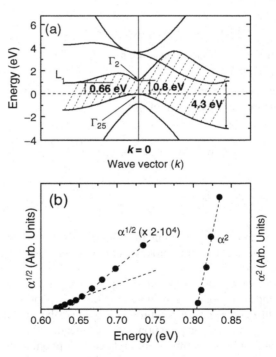

Figure 4.11   (a) The band structure of germanium (reproduced with permission from Cohen and Chelikowsky, 1998). (b) An analysis of the absorption edge of Ge. Notice the scale factor in the first plot in relation to the second plot (reproduced with permission from Dash and Newman, 1955).

The band structure of Ge, given in Figure 4.11(a), also shows a second band gap at 0.8 eV, which is now direct and corresponds to the $\Gamma_{25} \to \Gamma_2$ transition. Indeed, this direct band gap is also shown experimentally in the linear plot of α^2 versus the photon energy for energies larger than about 0.8 eV. According to the observed trend, $\alpha \propto (\hbar\omega - 0.8)^{1/2}$, we can say that these direct transitions are allowed (see Table 4.3).

Finally, we must say that (noting the scale factor in Figure 4.11) indirect absorption coefficients (from 0.6 to 0.75 eV for Ge) are in general insignificant compared to direct absorption coefficients (energies larger than 0.8 eV for Ge). This is because indirect absorption processes are much weaker than direct absorption processes, due to the second order nature of the former.

4.7 EXCITONS

The optical absorption of some semiconductors or insulator materials shows a series of peaks or features at photon energies close to but lower than the energy gap (the pre-edge region). These features correspond to a particular type of excitation called an *exciton*.

We know that the absorption of a photon with energy $\hbar\omega \geq \hbar\omega_g$ creates an electron in the conduction band and a hole in the valence band; both charges are free and can move independently in the crystal. However, because of the Coulomb attraction between these two opposite charges, an electron–hole pair can be created. This neutrally charged pair is called an *exciton*, and it can move through the crystal and transport energy without contributing to the electrical conductivity. As the electron and hole are bounded, excitons give rise to discrete levels at resonance energies close to but lower than the band-gap energy, as sketched in Figure 4.12. These energy levels are responsible for the *exciton transitions* observed below the energy gap.

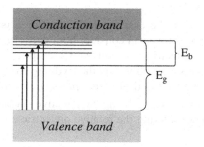

Figure 4.12 The energy levels of an exciton. The arrows indicate the possible optical exciton transitions in respect to the energy gap, E_g. E_b is the binding energy.

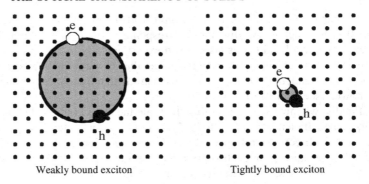

Weakly bound exciton   Tightly bound exciton

Figure 4.13   The schemes of (a) a weakly bound (Mott–Wannier) exciton and (b) a tightly bound (Frenkel) exciton.

The simplest exciton can be modeled by a hydrogen atom, with the electron and hole (equivalent to the hydrogen nucleus) in a stable orbit around each other. Within this model, two basic types of excitons can occur in crystalline materials:

- weakly bound (Mott–Wannier) excitons;
- tightly bound (Frenkel) excitons.

These two types of exciton are schematically illustrated in Figure 4.13. The *Mott–Wannier excitons* have a large radius in comparison to the interatomic distances (Figure 4.13(a)) and so they correspond to delocalized states. These excitons can move freely throughout the crystal. On the other hand, the *Frenkel excitons* are localized in the vicinity of an atomic site, and have a much smaller radius than the Mott–Wannier excitons. We will now describe the main characteristics of these two types of exciton separately.

4.7.1 Weakly Bound (Mott–Wannier) Excitons

For weakly bound excitons, the electron–hole separation is large compared to the interatomic distances, so that we can consider both particles as moving in a uniform crystal with a relative dielectric constant ε_r. Consequently, we can approach the energy levels of these excitons by utilizing the Rydberg energy levels of an hydrogen atom, but modified by the reduced effective mass, μ, of the electron–hole system ($1/\mu = 1/m_e^* + 1/m_h^*$, m_e^* and m_h^* being the effective masses of the electron and the hole, respectively) and considering the relative dielectric constant ε_r of the crystal.[2]

---

[2] We now use the symbol ε_r for the relative dielectric constant instead of the symbol ε that we used in Sections 4.2 and 4.3. In fact, the actual symbol is ε_r but, in the previous sections mentioned, we considered ε for the sake of simplicity in the different formulas.

Thus, the energy levels of such a hydrogen-like atom, measured from the ionization level (at 0 eV), are given by $-\left(\mu R_H / m_e \varepsilon_r^2\right)\left(1/n^2\right)$, where R_H is the Rydberg constant of the hydrogen atom (13.6 eV), m_e is the electronic mass, and n is the principal quantum number. Therefore, as the ionization level is at the bottom of the conduction band, the exciton energy levels in the crystal (see Figure 4.12) are given by

$$E_n = E_g - \frac{\mu R_H}{m_e \varepsilon_r^2} \frac{1}{n^2} \tag{4.39}$$

Thus, Mott–Wannier excitons can give rise to a number of absorption peaks in the pre-edge spectral region according to the different states $n = 1, 2, 3, \ldots$ As a relevant example, Figure 4.14 shows the low-temperature absorption spectrum of cuprous oxide, Cu_2O, where some of those hydrogen-like peaks of the excitons are clearly observed. These peaks correspond to different excitons states denoted by the quantum numbers $n = 2, 3, 4,$ and 5.

Within the simple Bohr model used for weakly bound excitons, the radius of the electron–hole orbit is given by

$$r_n = \frac{m_e \varepsilon_r a_B}{\mu} \times n^2 \tag{4.40}$$

where $a_B = 5.29 \times 10^{-11}$ m is the Bohr radius. According to Equations (4.39) and (4.40), the ground state $n = 1$ corresponds to the lowest energy and the shortest radius. Thus, the energy needed to ionize the exciton, called the *binding energy*, is given by

$$E_b = \frac{\mu R_H}{m_e \varepsilon_r^2} \tag{4.41}$$

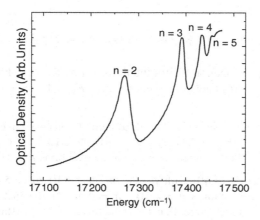

Figure 4.14   The absorption spectrum of cuprous oxide at 77 K, showing the exciton peaks $n = 2, 3, 4,$ and 5.

Table 4.4 The gap (E_g) and binding (E_b) energies of (Mott–Wannier and Frenkel) excitons in different materials

Crystal (Mott–Wannier excitons)	E_g (eV)	E_b (meV)	Crystal (Frenkel excitons)	E_g (eV)	E_b (meV)
GaN	3.5	23	LiF	13.7	900
ZnSe	2.8	20	NaF	11.5	800
CdS	2.6	28	KF	10.8	900
ZnTe	2.4	13	RbF	10.3	800
CdSe	1.8	15	NaCl	8.8	900
CdTe	1.6	12	KCl	8.7	900
GaAs	1.5	4.2	KBr	7.4	700
InP	1.4	4.8	KI	6.3	400
GaSb	0.8	2.0	NaI	5.9	300

For weakly bound (Mott–Wannier) excitons (mainly observed in semiconductors), the binding energies are in the meV range, as can be appreciated from Table 4.4. Inspection of this table also shows a general trend: E_b tends to increase as E_g increases. This is mainly because ε_r decreases and μ increases as the band gap increases.

EXAMPLE 4.5   *From the absorption spectrum in Figure 4.14, determine (a) the energy gap of Cu_2O, (b) the reduced effective mass of the exciton, assuming $\varepsilon_r = 10$, and (c) the Bohr radius of the exciton $n = 2$.*

(a) From Figure 4.14, we observe four well-defined exciton peaks at energies of 17 260 cm^{-1} ($n = 2$), 17 373 cm^{-1} ($n = 3$), 17 408 cm^{-1} ($n = 4$), and 17 426 cm^{-1} ($n = 5$). With these energies, we obtain the following empirical relation:

$$E(\text{cm}^{-1}) = 17\,458 - 800 \times (1/n^2)$$

According to Equation (4.39), the first term on the right-hand side of this expression corresponds to the energy gap: $E_g = 17\,458$ cm^{-1} = 2.16 eV, for Cu_2O.

(b) From the previous expression and Equation (4.39), we have $\mu R_H / m_e \varepsilon_r^2 = 800$ cm^{-1} = 0.099 eV. Thus, $\mu/m_e = \left(0.099 \times 10^2\right)/13.6 = 0.7$. We then find that the reduced effective mass of the exciton is $\mu = 0.7 m_e$.

(c) Using Equation (4.40) with $\mu = 0.7 m_e$, $\varepsilon_r = 10$, $a_B = 5.29 \times 10^{-11}$ m, and $n = 2$, we obtain $r_2 = 3 \times 10^{-9}$ m $= 3$ nm. Notice that this exciton radius (30 Å) is much larger than the average interatomic distances in crystals (of the order of Å), in accordance with the weakly bound nature of these excitons.

4.7.2 Tightly Bound (Frenkel) Excitons

In tightly bound (Frenkel) excitons, the observed peaks do not respond to the hydrogenic equation (4.39), because the excitation is localized in the close proximity of a single atom. Thus, the exciton radius is comparable to the interatomic spacing and, consequently, we cannot consider a continuous medium with a relative dielectric constant ε_r, as we did in the case of Mott–Wannier excitons.

Frenkel excitons are usually observed in large band-gap crystals. Highly illustrative examples occur in pure alkali halides, which are transparent crystals in the visible, with band gaps that lie in the ultraviolet spectral region. Figure 4.15 shows the room temperature absorption spectra of NaCl and LiF. The strong absorption peaks at 7.9 eV (NaCl) and 12.8 eV (LiF), close to the band gaps of these crystals, correspond to Frenkel excitons. The binding energies of these excitons are easily estimated, $E_b = E_g - E_{peak}$, giving about 0.9 eV for both NaCl and LiF crystals. In Table 4.4, the binding energies and band gaps of different alkali halides are given. An inspection of this table shows that binding energies are much larger than the Boltzmann energy at 300 K ($kT \approx 0.026$ eV), and therefore strong exciton peaks (see Figure 4.15) can be observed, even at room temperature. The exciton peaks in these crystals correspond to transitions localized on the anions (F^-, Cl^-, Br^-, and I^-), as they have lower electronic excitation levels than the cations. When the spectra are taken at cryogenic temperatures, additional structure can be observed on these peaks. This structure provides further information on the nature of the electronic energy levels of these excitons.

Frenkel excitons are also observed in many organic crystals and in noble gas crystals (Ne, Ar, Kr, and Xe). However, for the latter crystals the band gap lies out

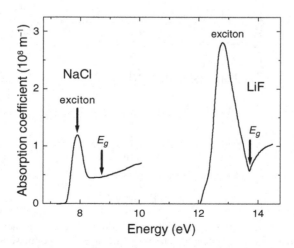

Figure 4.15 The absorption spectra (near the fundamental absorption edge) of sodium chloride and lithium fluoride at 300 K (reproduced with permission from Palik, 1985).

of the optical range (from 9.3 to 21.6 eV) and they only crystallize at cryogenic temperatures.

4.8 ADVANCED TOPIC: THE COLOR OF METALS

The simple free electron model (the Drude model) developed in Section 4.4 for metals successfully explains some general properties, such as the 'filter' action for UV radiation and their high reflectivity in the visible. However, in spite of the fact that metals are generally good mirrors, we perceive visually that gold has a yellowish color and copper has a reddish aspect, while silver does not present any particular color; that is it has a similarly high reflectivity across the whole visible spectrum. In order to account for some of these spectral differences, we have to discuss the nature of interband transitions in metals.

Figure 4.16 shows the reflectivity spectra of copper and silver. The plasma photon energies corresponding to Cu ($\hbar\omega_p = 10.8$ eV) and Ag ($\hbar\omega_p = 9$ eV), calculated

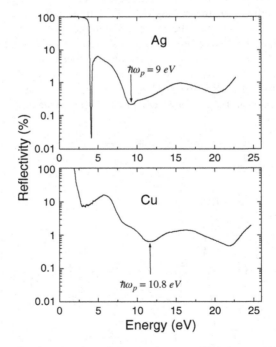

Figure 4.16   The reflectivity spectra of silver and copper. The photon energy corresponding to the plasma frequency is indicated in each case (reproduced with permission from Ehrenreich and Philipp, 1962).

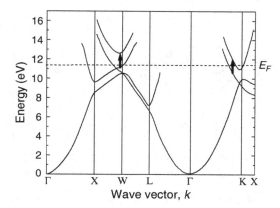

Figure 4.17   The band structure diagram of aluminum. The transitions in the vicinity of the W and K points, which are responsible for the reflectivity dip around 1.5 eV (see Figure 4.5), have been marked by arrows (reproduced with permission from Segall, 1961).

by Equation (4.20), do not explain the strong decrease in the reflectivity from about 2 eV (Cu) and 4 eV (Ag). These features are associated with interband absorption edges, which take into account both the band structure and the density of the states associated with initial and final energy bands. In fact, these features in the reflectivity spectra explain the reddish color of Cu and the colorless aspect of Ag.

To analyze these *interband effects*, we start by revisiting the reflectivity spectrum of aluminum, previously given in Figure 4.5. The Drude model fails to explain the dip at about 1.5 eV and the reduction in the reflectivity in respect to the free (or damped) carrier model. Figure 4.17 shows a band structure diagram for Al. The Fermi level energy, E_F, gives the limit between filled and empty states. As occurs for direct transitions in insulators and semiconductors (see Equation (4.31)), the absorption coefficient in metals is proportional to the density of states of the transitions involved. Thus, the dip in the reflectivity at about 1.5 eV (Figure 4.5) is related to transitions in the vicinity of the W and K points in the band structure diagram of Al, because of the so-called *parallel band effect*. This effect occurs because two bands (above and below the Fermi level) are nearly parallel and then a large number of transitions can occur at almost the same photon energy, 1.5 eV. In other words, it means that the density of states at this energy is very high, leading to a strong absorption (a strong dip in reflectivity).

By a careful inspection of Figure 4.17, we see how further transitions between bands below and above the Fermi level can also occur at energies higher than 1.5 eV. However, as these bands are not parallel, the density of states at these energies is lower than at 1.5 eV. In any case, the absorption probability is still significant, and it accounts for the experimentally observed reduction in the reflectivity of Al in respect to the predictions from the Drude model (see Figure 4.5).

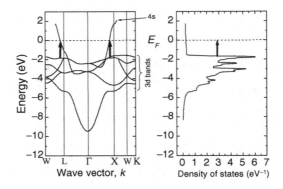

Figure 4.18   The band structure of Cu (left-hand side): the two arrows indicate the transitions from the 3d bands that are responsible for the decrease in reflectivity around 2 eV (see Figure 4.16). The density of states of Cu (right-hand side) as calculated from its band structure. The arrows indicate the threshold for 3d → 4s transitions (reproduced with permission from Moruzzi *et al.*, 1978).

The interpretation of the spectral features at frequencies lower than the plasma frequency for Cu and Ag (Figure 4.16) is a little bit more complicated than for Al. We will focus our attention on the case of copper (Cu); a metal with an outer electronic configuration of $3d^{10}4s^1$. Figure 4.18 shows the band structure and the density of states of Cu. The sharp peaks in the density of states are essentially related to 3d electrons, which lie below the Fermi level, E_F. On the other hand, the 4s band is only half-filled; that is, the Fermi level lies in the middle of this band. Then 3d → 4s interband transitions can be clearly observed. The lowest energy transitions among the 3d → 4s transitions, indicated by arrows in Figure 4.18, occur at about 2 eV. This explains the decrease in the reflectivity spectra observed at about this energy for metallic Cu (Figure 4.16, bottom), and hence the reddish aspect of copper. However, the threshold for 3d → 4s transitions of Ag occurs around 4 eV. This explains the strong dip in the reflectivity spectrum of this metal at about this energy (see Figure 4.16, top), and hence the colorless aspect of silver, as all of the visible spectral region is 100 % reflected.

EXERCISES

4.1 The sodium chloride (NaCl) crystal shows very high absorption and reflectivity in the infrared region, known as the 'Restrahlen region.' The real and imaginary relative dielectric constants at 6000 nm are, respectively, $\varepsilon_1 = 16.8$ and $\varepsilon_2 = 91.4$. At this wavelength, estimate (a) the refractive index and the extinction coefficient, and (b) the optical density and the reflectivity at normal incidence for a 1 mm thick NaCl sample. (c) If the previous sample is illuminated (at normal incidence) by a beam of intensity I_0 at 6000 nm, estimate the intensity of this beam after it passes through the sample. (Take into account both the absorption and reflection processes at the front and back surfaces of the sample.)

4.2 Sodium (Na) gas shows two well-defined absorption bands that peak at 589 nm and 589.6 nm. Consider this gas as a diluted medium (so that $n \approx 1$), with a density of ions of $N = 1 \times 10^{11}$ cm^{-3} and a damping rate of $\Gamma = 628$ MHz. (a) Estimate the absorption coefficient at these two peaks. (b) Now estimate the power of a laser beam of 1 mW at 589.0 nm after it passes through a Na cell of thickness 5 cm.

4.3 Figure E4.3 shows the room temperature absorption spectra of an insulator (LiNbO$_3$), a semiconductor (Si), and a metal (Cu). (a) Determine the spectrum associated with each one of these materials. (b) From these spectra, estimate the energy-gap values of Si and LiNbO$_3$ and the plasma frequency of Cu. (c) What can be said about the transparency in the visible range for each of these materials?

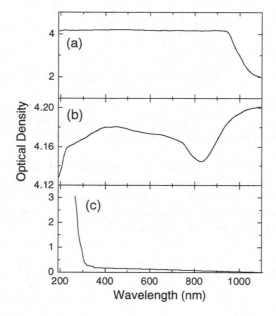

Figure E4.3   The room temperature absorption spectra of three solid materials (see the text in Exercise 4.3).

4.4 Zinc (Zn) is a divalent metal with an atomic density of 6.6×10^{22} cm^{-3}. Determine the wavelength region within the optical range for which you expect this metal to be a good mirror.

4.5 Let us consider an ideal metal with 2×10^{22} cm^{-3} valence electrons. For such a metal, estimate (a) its plasma frequency, (b) the optical density at 400 nm for a sample of thickness 0.1 mm, and (c) the reflectivity at 300 nm. Will the metal be transparent at this wavelength?

4.6 Figure E4.6 (left-hand side) shows the band structure of cadmium telluride (CdTe), where the shadowed region corresponds to the 'gap' of this semiconductor. (a) On the basis of this diagram, do you expect CdTe to be a direct or an indirect material? (b) Using Figure E4.6 (right-hand side), estimate the energy gap and determine whether the interband transitions are allowed or forbidden. (c) Do you expect CdTe to be transparent in the visible?

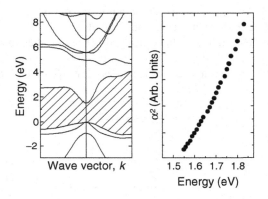

Figure E4.6 (a) The band structure of CdTe (left) and the square of the absorption coefficient versus photon energy for CdTe (right).

4.7 The analyzed optical absorption spectrum of a given thin-film semiconductor sample is displayed in Figure E4.7. (a) From this figure, estimate the energy gap of such a semiconductor and infer its direct or indirect nature. (b) Sketch a possible band structure diagram for this semiconductor, indicating the main transitions that you infer from Figure E4.7.

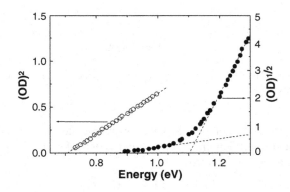

Figure E4.7 The absorption spectrum of a thin-film semiconductor sample (OD stands for optical density).

4.8 Indium phosphide (InP) has a band-gap energy of 1.424 eV. For this semiconductor $m_e^* = 0.077m_e$, $m_h^* = 0.2m_e$, and $\varepsilon_r = 12.4$. (a) Estimate the wavelength peaks associated with the $n = 1$, $n = 2$, and $n = 3$ states of the weakly bound (Mott–Wannier) excitons. (b) Determine the binding energies of these excitons. (c) Calculate the energy required to promote an exciton from the ground state, $n = 1$, to the next higher excited state, $n = 2$ ($R_H = 13.6$ eV).

4.9 The absorption spectrum of ultrapure gallium arsenide (GaAs) at 1.2 K shows three sharp absorption peaks at 1.5149 eV, 1.5180 eV, and 1.5187 eV, corresponding to the exciton states $n = 1$, $n = 2$, and $n = 3$, respectively. Determine (a) the energy band gap of GaAs, and (b) the reduced mass of the exciton, considering that $\varepsilon_r = 12.8$. (c) GaAs has a cubic structure with a unit cell of 0.56 nm. Estimate the number of units cells within the orbit of the $n = 1$ exciton. (d) Estimate the highest temperature for which it is possible to observe exciton absorption peaks.

REFERENCES AND FURTHER READING

Electronic Structure and Optical Properties of Semiconductors, Cohen, M. L. and Chelikowsky, J., Springer-Verlag, Berlin (1988).

Dash, W. C. and R., Newman, *Phys. Rev.*, **99**, 1151 (1955).

Ehrenreich, H., Philipp, H. R., and Segall, B., *Phys. Rev.*, **132**, 1918 (1962).

Ehrenreich, H. and Philipp, R. H., *Phys. Rev.*, **128**(4), 1622 (1962).

Fox, M., *Optical Properties of Solids*, Oxford University Press (2001).

Henderson, B. and Imbusch, G. F., *Optical Spectroscopy of Inorganic Solids*, Oxford Science Publications (1989).

Moruzzi, V. L., Janak, J. F., and Williams, A. R., *Calculated Electronic Properties of Metals*, Pergamon Press, New York (1978).

Palik, E. D. (ed.), *Handbook of Optical Constants of Solids*, Academic Press, San Diego (1985).

Philipp, H. R. and Ehrenreich, H., *Phys. Rev.*, **129**, 1550 (1963a).

Philipp, H. R. and Ehrenreich, H., *Phys. Rev.*, **131**, 2016 (1963b).

Segall, B., *Phys. Rev.*, 124, 1797 (1961).

Svelto, O., *Principles of Lasers*, Plenum Press (1986).

Weber, M. J., *Handbook of Optical Materials*, CRC Press, Boca Raton, Florida (2003).

Wooten, F., *Optical Properties of Solids*, Academic Press, New York (1972).

Yu, P. Y. and Cardona, M., *Fundamentals of Semiconductors. Physics and Materials Properties*, Springer-Verlag, Berlin (1999).

5

Optically Active Centers

5.1 INTRODUCTION

A variety of interesting optical properties and applications of inorganic materials depend on the presence of so-called *optically active centers*. These centers consist of dopant ions that are intentionally introduced into the crystal during the growth process, or lattice defects (color centers) that are created by various methods. Both types of localized center provide energy levels within the energy gap of the material, so that they can give rise to the appearance of optical transitions at frequencies lower than that of the fundamental absorption edge. In this chapter, we shall mostly deal with centers due to dopant ions. However, as will be shown in the next chapter, the content of this chapter can also apply to color centers.

The optical features of a center depend on the type of dopant, as well as on the lattice in which it is incorporated. For instance, Cr^{3+} ions in Al_2O_3 crystals (the ruby laser) lead to sharp emission lines at 694.3 nm and 692.8 nm. However, the incorporation of the same ions into $BeAl_2O_4$ (the alexandrite laser) produces a broad emission band centered around 700 nm, which is used to generate tunable laser radiation in a broad red–infrared spectral range.

Instead of considering how the incorporation of a dopant ion perturbs the electronic structure of the crystal, we will face the problem of understanding the optical features of a center by considering the energy levels of the dopant free ion (i.e., out of the crystal) and its local environment. In particular, we shall start by considering the energy levels of the dopant free ion and how these levels are affected by the presence of the next nearest neighbors in the lattice (the environment). In such a way, we can practically reduce our system to a one-body problem.

Let us consider a dopant ion A (the *central ion*) placed at a lattice site, surrounded by an array of six regular lattice ions B (*ligand ions*), separated by a distance a from the ion A. The ligand ions B are located at the corners of an octahedron, as shown in

An Introduction to the Optical Spectroscopy of Inorganic Solids   J. García Solé, L. E. Bausá, and D. Jaque
© 2005 John Wiley & Sons, Ltd   ISBNs: 0-470-86885-6 (HB); 0-470-86886-4 (PB)

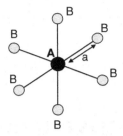

Figure 5.1 A scheme of an illustrative optical center, AB_6. This particular center consists of a dopant optical ion A in an octahedral environment of B ions.

Figure 5.1. The set comprised of ion A and the six ligand ions B constitutes a pseudo-molecule, AB_6, which we call a *center*. This is a common arrangement (center) for optical ions in solids, called an octahedral arrangement, and we will use it in the text as our reference center. Of course, many other arrangements around ion A are possible, but the strategy for solving each particular case is quite similar to the one for our selected octahedral center.

In this chapter, we will treat those centers that produce the appearance of optical bands. This type of center is called an *optically active center*. We will try to understand how these centers give rise to the appearance of new optical bands (which are not present in the undoped crystal) and to predict their main features (spectral location, intensity, shape, etc.).

5.2 STATIC INTERACTION

We will now consider our illustrative optical center AB_6 (Figure 5.1) in order to determine its energy levels. Let us assume the ideal situation of a rigid (nonvibrating) lattice, so that the A–B distance a remains fixed and corresponds to the time-average positions in the vibrating case. We know that the energy levels of A are modified (shifted and split) by the influence of the ligand ions B through the electric field that they produce at the site of A. This static electric field is commonly called the *crystalline field*.

In order to account for the optical absorption and emission bands of the AB_6 center, we must first determine its energy levels E_i by solving the Schrödinger equation:

$$H\psi_i = E_i\psi_i \tag{5.1}$$

where H represents a Hamiltonian that includes the different interactions of the valence electrons in the AB_6 center and ψ_i are eigenfunctions of the center. Depending on the particular type of center, two main methodologies are commonly used to solve the Schrödinger equation (5.1): *crystalline field theory* and *molecular orbital theory*. We will discuss them separately.

5.2.1 Crystalline Field Theory

In crystalline field theory, the valence electrons belong to ion A and the effect of the lattice is considered through the electrostatic field created by the surrounding B ions at the A position. This electrostatic field is called *the crystalline field*. It is then assumed that the valence electrons are localized in ion A and that the charge of B ions does not penetrate into the region occupied by these valence electrons. Thus the Hamiltonian can be written as

$$H = H_{FI} + H_{CF} \tag{5.2}$$

where H_{FI} is the *Hamiltonian* related to the *free ion* A (an ideal situation in which the A ions are isolated, similar to a gas phase of these ions) and H_{CF} is the *crystal field Hamiltonian*, which accounts for the interaction of the valence electrons of A with the electrostatic crystal field created by the B ions. The crystal field Hamiltonian can be written as

$$H_{CF} = \sum_{i=1}^{N} eV(r_i, \theta_i, \varphi_i) \tag{5.3}$$

where $eV(r_i, \theta_i, \varphi_i)$ is the potential energy created by the six B ions at the position $(r_i, \theta_i, \varphi_i)$ (given in spherical coordinates) of the ith valence electron of ion A. The summation is extended over the full number of valence electrons (N).

In order to apply quantum mechanical perturbation theory, the free ion term is usually written as

$$H_{FI} = H_0 + H_{ee} + H_{SO} \tag{5.4}$$

where H_0 is the central field Hamiltonian (a term that reflects the electric field acting on the valence electrons due to the nucleus and the inner- and outer-shell electrons), H_{ee} is a term that takes into account any perturbation due to the Coulomb interactions among the outer (valence) electrons, and H_{SO} represents the spin–orbit interaction summed over these electrons.

Depending upon the size of the crystal field term H_{CF} in comparison to these three free ion terms, different approaches can be considered to the solution of Equation (5.1) by perturbation methods:

- *Weak crystalline field*: $H_{CF} \ll H_{SO}, H_{ee}, H_0$. In this case, the energy levels of the free ion A are only slightly perturbed (shifted and split) by the crystalline field. The free ion wavefunctions are then used as basis functions to apply perturbation theory, H_{CF} being the perturbation Hamiltonian over the $^{2S+1}L_J$ states (where S and L are the spin and orbital angular momenta and $J = L + S$). This approach is generally applied to describe the energy levels of trivalent rare earth ions, since for these ions the 4f valence electrons are screened by the outer $5s^2\, 5p^6$ electrons. These electrons partially shield the crystalline field created by the B ions (see Section 6.2).

- *Intermediate crystalline field*: $H_{SO} \ll H_{CF} < H_{ee}$. In this case, the crystalline field is stronger than the spin–orbit interaction, but it is still less important than the interaction between the valence electrons. Here, the crystalline field is considered a perturbation on the ^{2S+1}L terms. This approach is applied for transition metal ion centers in some crystals (see Section 6.4).

- *Strong crystalline field*: $H_{SO} < H_{ee} < H_{CF}$. In this approach, the crystalline field term dominates over both the spin–orbit and the electron–electron interactions. This applies to transition metal ions in some crystalline environments (see Section 6.4).

To illustrate how the perturbation problem must be solved, we now describe one of the simplest cases, corresponding to an octahedral crystalline field acting on a single d^1 valence electron.

The crystalline field on d^1 optical ions

One of the simplest descriptions of the crystalline field occurs for the d^1 outer electronic configuration (i.e., for a single d^1 valence electron). This means that $H_{ee} = 0$ and, consequently, there is no distinction between intermediate and strong crystalline fields.

Let us assume this outer electronic configuration for the ion A of our AB_6 center (Figure 5.1); that is, a d^1 electron in an octahedral crystalline field. This could correspond, for instance, to the case of Ti^{3+} ions ($3d^1$ outer electronic configuration) in Al_2O_3, a crystal called Ti–sapphire, which is used for broadly tunable solid state lasers (see Section 2.5). In this crystal, the Ti^{3+} ion (ion A) is surrounded by six O^{2-} ions (B ions). Although the actual environmental symmetry is slightly distorted from that of Figure 5.1, we can adopt this octahedral surrounding as a first order approximation.

For free Ti^{3+} ions – that is, in the absence of the crystalline field – the Hamiltonian has spherical symmetry and the angular eigenfunctions of the $3d^1$ states are the spherical harmonics, $Y_l^{m_l}$ (given in Appendix A2), with $l = 2$ and $m_l = 2, 1, 0, -2, -1$. Therefore, the $3d^1$ state is fivefold degenerate.

In Appendix A2, we have formally applied the perturbation method to find the energy levels of a d^1 ion in an octahedral environment, considering the ligand ions as point charges. However, in order to understand the effect of the crystalline field over d^1 ions, it is very illustrative to consider another set of basis functions, the d orbitals displayed in Figure 5.2. These orbitals are real functions that are derived from the following linear combinations of the spherical harmonics:

$$d_{z^2} \alpha Y_2^0, \qquad d_{x^2-y^2} \alpha \left(Y_2^2 + Y_2^{-2}\right)$$
$$d_{xy} \alpha - i \left(Y_2^2 - Y_2^{-2}\right), \qquad d_{xz} \alpha - \left(Y_2^1 - Y_2^{-1}\right),$$
$$d_{yz} \alpha \, i \left(Y_2^1 + Y_2^{-1}\right) \tag{5.5}$$

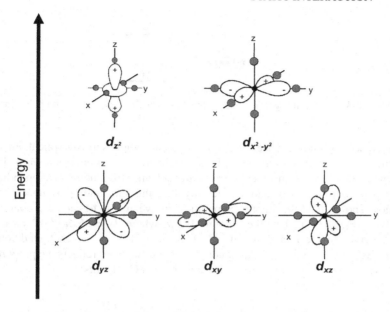

Figure 5.2 The effect of an octahedral arrangement of ligand ions (for instance, O^{2-} ions) on the d orbitals of a central ion (for instance, the Ti^{3+} ion). The shaded circles represent the ligand oxygen ions (B in Figure 5.1). The black point represents the central ion (A in Figure 5.1).

Figure 5.2 shows that, due to its similar symmetry, three d^1 orbitals (d_{xy}, d_{xz}, and d_{yz}) of the central ion (for instance, the Ti^{3+} ion) are affected in the same way by the octahedral environment of the ligand ions (for instance, the O^{2-} ions); these orbitals will have the same energy. Although it is not so intuitive, it can also be shown that orbitals d_{z^2} and $d_{x^2-y^2}$ are affected in a similar way by the octahedral environment of the ligand ions, then having equal energy. This means that the d fivefold degenerate energy state splits into two energy levels in an octahedral environment; a triply degenerate one, associated with the d_{xy}, d_{xz}, and d_{yz} orbitals (called t_{2g}), and another doubly degenerate one, associated with the d_{z^2} and $d_{x^2-y^2}$ orbitals (called e_g). The nomenclature used for labeling the crystalline field split levels is based on group theory considerations, which are treated in Chapter 7.

By further inspection of Figure 5.2, we can appreciate that the lobes of the d_{xy}, d_{zx}, and d_{yz} orbitals are accommodated between oxygen ions. This produces a more stable situation (lower energy) than that for the d_{z^2} and $d_{x^2-y^2}$ orbitals, for which the lobes always point toward oxygen ions. This aspect suggests that the t_{2g} energy level must lie below the e_g energy level, as shown in Figure 5.3.

The absorption spectrum of Ti^{3+} ions in Al_2O_3 (shown in Figure 5.4) provides experimental evidence of the previously discussed splitting of d energy levels: the broad band at around 500 nm is a due to the $t_{2g} \rightarrow e_g$ transition.

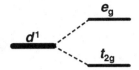

Figure 5.3   The splitting of a d energy level as a result of an octahedral environment.

Our discussion for the case of Ti^{3+} in Al_2O_3 also applies to any optical ion A with an outer d^1 electronic configuration in an octahedral crystal environment of B ions. A formal calculation of the energy levels for such an AB_6 center (see Appendix A2), using perturbation theory and considering that the surrounding B ions are point charges, shows that the energy separation between the t_{2g} and e_g energy levels ($E_{e_g} - E_{t_{2g}}$) is equal to the amount $10Dq$, where $D = 35Ze^2/4a^5$ is a factor that depends on the ligand B ions (Ze being the charge of each ligand ion) and $q = (2/105)\langle r^4 \rangle$ (r being the radial position of the electron) reflects the properties of the d^1 valence electron. Therefore, we can write (in CGS units)

$$E_{e_g} - E_{t_{2g}} = 10Dq = \frac{10}{6} Z \frac{e^2 \langle r^4 \rangle}{a^5} \tag{5.6}$$

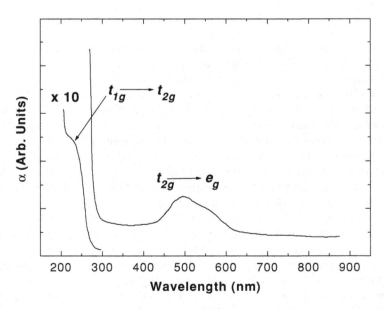

Figure 5.4   The absorption spectrum of Ti^{3+} in Al_2O_3 at room temperature.

EXAMPLE 5.1   *The estimation of 10Dq.*

Using Equation (5.6), the Dq parameter can be estimated for a given d^1 optical ion in an octahedral environment of Ze charged ligand ions at a distance a from the central ion, provided that the $\langle r^4 \rangle$ term is known. Although this term is difficult to calculate in many cases, we can make a rough estimation of the crystalline-field $10Dq$ by considering $r \approx 2a_0$, where $a_0 \approx 0.5$ Å is the hydrogen Bohr radius. Consequently, $\langle r^4 \rangle \approx r^4 = 16a_0^4$. We also assume a typical A–B distance of $a \approx 5a_0 \approx 2.5$ Å and a ligand point charge $2e$; that is, with $Z = 2$ (as for the O^{2-} ligand ions in the Ti–sapphire system).

Substituting these values in Equation (5.6), we obtain:

$$10Dq = \frac{10}{6} \times 2 \times \frac{e^2 \times 16}{5^5 \times a_0} = \frac{10}{6} \times 2 \times \frac{(1.6 \times 10^{-19} \times 3 \times 10^9 \text{ stc})^2 \times 16}{5^5 \times 0.5 \times 10^{-8} \text{ cm}}$$

$$= 7.8 \times 10^{-13} \text{ erg} \approx 4000 \text{ cm}^{-1}$$

This value is far from those experimentally obtained from the absorption spectra of d^1 ions in octahedral fields (which range from about $10\,000$ cm^{-1} to about $20\,000$ cm^{-1}). Our estimated $10Dq$ value corresponds to an infrared absorption wavelength of 2500 nm, far from the observed absorption wavelength peak at 500 nm, corresponding to the $t_{2g} \rightarrow e_g$ transition of Ti^{3+} in Al_2O_3 (see Figure 5.4).

The accordance between the experimental and calculated Dq values can be improved slightly if the actual nonpoint charge distribution of B ions and covalence effects are considered. However, the Dq value is usually obtained from experimental measurements, and this value is considered to be an empirical parameter, as the dependency on a (the distance A–B) is explained well.

So far, we have discussed the crystalline field acting on the ion A due to an octahedral environment of six B ligand ions. In many optically ion activated crystals, such as $Ti^{3+}:Al_2O_3$, the local symmetry of the active ion A is slightly distorted from the perfect octahedral symmetry (O_h symmetry). This distortion can be considered as a perturbation of the main octahedral field. In general, this perturbation lifts the orbital degeneracy of the t_{2g} and e_g levels and then produces additional structure in the $t_{2g} \leftrightarrow e_g$ absorption/emission bands.

On the other hand, the crystalline field due to main symmetries other than O_h symmetry can be also related to this same case. For this purpose, it is useful to represent the octahedral structure of our reference AB_6 center as in Figure 5.5(a). In this representation, the B ions lie in the center of the six faces of a regular cube of side $2a$ and the ion A (not displayed in the figure) is in the cube center; the distance A–B is equal to a.

The advantage of this representation is that other typical arrangements can also be displayed using this regular cube, as shown in Figures 5.5(b) and 5.5(c). The

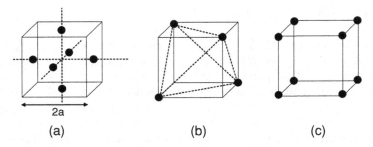

2a

(a)   (b)   (c)

Figure 5.5 The arrangement of B ligand ions (black dots) around a central ion A (located in the cube center, but not displayed in the figure): (a) octahedral, (b) tetrahedral, and (c) cubic.

arrangement in Figure 5.5(b) corresponds to an AB_4 center with a tetrahedral structure. This arrangement (T_d symmetry) consists of four ligand B ions located at the alternate vertices of the regular cube – that is, at the vertices of a regular tetrahedron – while the ion A lies at the cube center. If we preserve the cube side as $2a$, the distance A–B is now $a\sqrt{3}$. The arrangement of Figure 5.5(c) corresponds to an AB_8 center with cubic symmetry (O_h symmetry). This center can be visualized by locating the eight B ligands at the cube corners, with the central ion A at the cube center. Again the distance A–B is $a\sqrt{3}$.

The regular cube used in Figure 5.5 to represent different symmetry centers suggests that these symmetries can be easily interrelated. In particular, following the same steps as in Appendix A2, it can be shown that the crystal field strengths, $10Dq$, of the tetrahedral and cubic symmetries are related to that of the octahedral symmetry. Assuming the same distance A–B for all three symmetries, the relationships between the crystalline field strengths are as follows (Henderson and Imbusch, 1989):

$$Dq(\text{octahedral}) = -\tfrac{9}{4}Dq(\text{tetrahedral}) = -\tfrac{9}{8}Dq(\text{cubic}) \tag{5.7}$$

We then see that the crystalline field splitting in octahedral symmetry is larger by a factor of 9/4 than in tetrahedral symmetry, and larger by a factor of 9/8 than in cubic symmetry. The minus sign indicates an inversion of the e_g and t_{2g} levels with respect to the octahedral field, as shown in Figure 5.6. Therefore, all calculations of the crystalline field splitting carried out for the AB_6 center can be used, with the appropriate changes in the $10Dq$ values, for the AB_4 and AB_8 centers displayed in Figures 5.5(b) and 5.5(c).

Finally, we must say that the calculation of the crystalline field splitting for multi-electron d^n states is much more complicated than for d^1 states. For d^n states ($n > 1$), electrostatic interactions among the d electrons must be taken into account, together with the interactions of these valence electrons with the crystalline field.

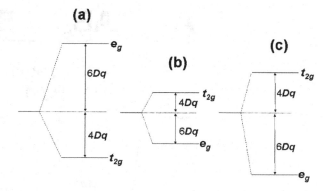

Figure 5.6   The energy-level scheme for the crystal field splitting of a d^1 electron in different symmetries: (a) octahedral, (b) tetrahedral, and (c) cubic.

5.2.2 Molecular Orbital Theory

Molecular orbital theory is a semi-empirical method devoted to interpreting the energy-level structure of optical centers where the valence electron cannot be considered as belonging to a specific ion. In our AB_6 reference center, this would mean that the valence electrons are shared by A and B ions. The approach is based on the calculation of molecular orbitals (MO) of the AB_6 pseudo-molecule, ψ_{MO}, from various trial combinations of the individual atomic orbitals, ψ_A and ψ_B, of the A and B ions, respectively. The molecular orbitals ψ_{MO} of the center AB_6 are conveniently written in the form

$$\psi_{MO} = N(\psi_A + \lambda \psi_B) \tag{5.8}$$

where N is a normalized constant and λ is a mixing coefficient. These MO functions are approximate solutions to the Schrödinger equation (5.1), and their validity as wavefunctions for the AB_6 center is tested by calculating observable quantities with these MO functions and then comparing with experimental results.

Figure 5.7 shows the MO energy-level scheme for an octahedral AB_6 center, A being a transition metal cationic ion and B the anionic ions. The different MO and energy levels of the molecule are determined from the outer atomic orbitals 3d, 4s, and 4p of the ions A (the left-hand side in Figure 5.7) and the p and s atomic orbitals of the ligand B ions (the right-hand side in Figure 5.7). Due to symmetry reasons, only linear combinations of the ligand s and p orbitals must be considered for bonding, giving rise to the $\pi(p_x, p_y)$ and $\sigma(s - p_z)$ energy levels of B in Figure 5.7. The resulting MO energy levels of the AB_6 center are denoted by specific labels related to symmetry properties followed by the type of spatial bonding, (σ) or (π), between the A and B

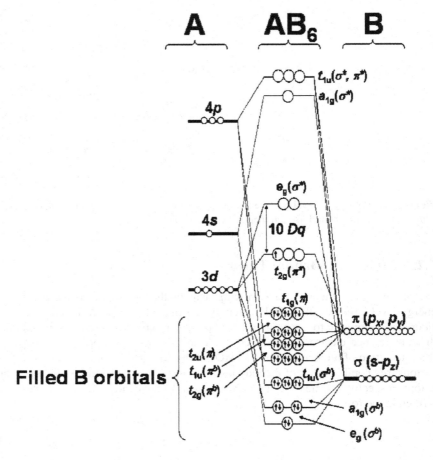

Figure 5.7 A schematic energy-level diagram for an octahedral AB_6 center within molecular orbital theory. This diagram is constructed from the atomic levels of A and B. The filled and half-filled states (two possible opposite spins for each state) correspond to A = Ti^{3+} and B = O^{2-} ions (reproduced with permission from Ballhausen and Gray, 1965).

atomic orbitals.[1] Each type of bonding produces a favorable energetic situation, σ or π *bonding*, and an unfavorable energetic situation, $\sigma*$ or $\pi*$ *antibonding*. The antibonding states correspond to the higher energy levels of the AB_6 center and are identified by an asterisk. The bonding states correspond to the lower energy levels of the AB_6 center.

---

[1] A σ bonding can occur between s orbitals or p orbitals, while π spatial bonding occurs only between p orbitals.

Let us consider an octahedral center $Ti^{3+}O_6^{2-}$; for instance, the absorbing center in the Ti–sapphire crystal. The outer electronic configuration of the Ti^{3+} ions is $3d^1$, while each O^{2-} ion has an outer electronic configuration $2p^6$. Thus, there are 36 electrons given by the six O^{2-} ions (B ions) which fill the lower energy levels of the AB_6 center. For symmetry reasons, the $2p_z$ atomic orbitals of the O^{2-} ions are combined with the 3s atomic orbitals of the O^{2-} ions to provide σ bonds, while the $2p_x$ and $2p_y$ atomic orbitals provide π bonds. As the central Ti^{3+} ion provides a 3d electron in the filling of the MO energy levels, the $Ti^{3+}O_6^{2-}$ energy levels are populated as in Figure 5.7, where each arrow indicates a spin direction. Therefore the last nonempty (partially filled) energy level of the AB_6 center is the antibonding $t_{2g}(\pi^*)$ level. In fact, the Ti–sapphire absorption band around 500 nm (Figure 5.4) corresponds to the $t_{2g}(\pi^*) \rightarrow e_g(\sigma^*)$ transition, the same transition predicted by crystalline field theory, which gives the crystal field strength, $10Dq$. However, higher absorption bands can arise from the $t_{1g}(\pi) \rightarrow t_{2g}(\pi^*)$ and $t_{1g}(\pi) \rightarrow e_g(\sigma^*)$ transitions, which are only explained by MO theory. In fact, the shoulder near 240 nm that appears in the absorption spectrum of Figure 5.4 must be attributed to one of those transitions, the $t_{1g}(\pi) \rightarrow t_{2g}(\pi^*)$ transition. These high-energy transitions promote electrons that mainly belong to states of the O^{2-} ligand ions to states that mainly belong to the Ti^{3+} ion, and so are called *charge transfer transitions*.

Thus, MO theory is generally applied to interpretation of the so-called *charge transfer spectra*. However, for a great variety of centers in solids, crystalline field theory suffices to provide at least a qualitative interpretation of spectra.

5.3 BAND INTENSITIES

In the previous section we have seen how to determine the energy levels of an optically active center. Optical spectra result from transitions among these energy levels. For instance, an optical absorption spectrum is due to different transitions between the ground energy level and the different excited energy levels. The absorption coefficient at each wavelength is proportional to the transition probability of the related transition.

In this section, we will study the absorption and emission probabilities for a single two-level atomic center that is illuminated by a monochromatic electromagnetic wave.

5.3.1 The Absorption Probability

We know from Chapter 1 that the probability P_{if} of inducing an optical transition from a state i to a state f is proportional to $|\langle \psi_f | H | \psi_i \rangle|^2$, where in the matrix element Ψ_i and Ψ_f denote the eigenfunctions of the ground and excited states, respectively, and H is the interaction Hamiltonian between the incoming light and the system (i.e., the valence electrons of the center). In general, we can assume that H is a sinusoidal function of time with a frequency ω equal to that of the incident wave. Then,

$$H = H^0 \sin \omega t \tag{5.9}$$

The next step is to apply basic time-dependent perturbation theory to our simple two-level center that is subjected to this time-varying interaction. After solving this basic problem (Svelto, 1986), the transition probability P_{if} is given by

$$P_{if} = \frac{\pi}{2\hbar^2} |H_{if}^0|^2 \delta(\Delta\omega) \tag{5.10}$$

where $H_{if}^0 = \langle \psi_f | H^0 | \psi_i \rangle$ and $\delta(\Delta\omega) = \delta(\omega - \omega_0)$ indicates that the transition is only possible for incident monochromatic radiation of appropriate frequency $\omega = \omega_0$. This Dirac δ-function is physically unacceptable, as the optical bands show defined shapes. Therefore, it should be replaced by the corresponding line-shape function $g(\omega)$, as we did for the transition cross section in Section 1.3.

If the transition is of an electric dipole nature, the interaction Hamiltonian can be written as $H = \mathbf{p} \cdot \mathbf{E}$, where $\mathbf{p}$ is the electric dipole moment and $\mathbf{E}$ is the electric field of the radiation. The electric dipole moment is given by $\mathbf{p} = \sum_i e\mathbf{r}_i$, where $\mathbf{r}_i$ is the position of the ith valence electron (measured from the nucleus of our center) and the summation is over all of the valence electrons. Normally, only one electron change of state is considered in the optical transition, so that $\mathbf{r}_i = \mathbf{r}$ and $\mathbf{E} = \mathbf{E}(\mathbf{r}, t)$.

Now we assume that the wavelength of the electromagnetic wave is much larger than the atomic dimensions. This is, of course, true for the optical range, as the shortest wavelength is around 200 nm while the atomic dimensions are of the order of 0.1 nm. In this case, the electric field does not vary within the atomic volume and so $\mathbf{E} \cong \mathbf{E}(0, t) = \mathbf{E}_0 \sin \omega t$. Therefore, we can write

$$H_{if}^0 = \mathbf{E}_0 \cdot \boldsymbol{\mu}_{if} \tag{5.11}$$

where $\mathbf{E}_0$ is the value at the nucleus and

$$\boldsymbol{\mu}_{if} = \langle \psi_f | e\mathbf{r} | \psi_i \rangle = e \int \psi_f^* \, \mathbf{r} \, \psi_i \, \mathrm{d}V \tag{5.12}$$

is the so-called *matrix element of the electric dipole moment*. If θ is the angle between $\mathbf{E}_0$ and $\boldsymbol{\mu}_{if}$, the squared matrix element of Equation (5.10) becomes

$$|H_{if}^0|^2 = E_0^2 |\boldsymbol{\mu}_{if}|^2 \cos^2 \theta \tag{5.13}$$

where $|\boldsymbol{\mu}_{if}| = \boldsymbol{\mu}_{if}^* \cdot \boldsymbol{\mu}_{if}$, since $\boldsymbol{\mu}_{if}$, is a complex vector.

Assuming now that the incident wave interacts with centers whose $\boldsymbol{\mu}_{if}$ vectors are randomly oriented with respect to $\mathbf{E}_0$, we can average Equation (5.13) over all possible orientations. Taking into account that $\langle \cos^2 \theta \rangle = 1/3$ (considering that all orientations θ are equally probable), we obtain $\langle|H_{if}^0|^2\rangle = \frac{1}{3}E_0^2|\mu_{if}|^2$. We can then rewrite expression (5.10) and give a more detailed expression for the absorption probability of our two-level center, as follows:

$$P_{if} = \frac{\pi}{3n\varepsilon_0 c_0 \hbar^2} I |\mu_{if}|^2 \delta(\Delta\omega) \qquad (5.14)$$

where $I = \frac{1}{2}nc_0\varepsilon_0 E_0^2$ is the intensity of the incident radiation (assuming an incident plane wave), c_0 is the speed of light in a vacuum, n is the refractive index of the absorbing medium, and ε_0 is the permittivity in a vacuum.

Expression (5.14) shows that the absorption probability depends on both the incoming light intensity and the matrix element μ_{if}. It is easy to see that $|\mu_{if}| = |\mu_{fi}| = |\mu|$ and so we can conclude that the absorption probability between two defined energy levels i and f is equal to the stimulated emission probability between levels f and i:

$$P_{if} = P_{fi} = P \qquad (5.15)$$

5.3.2 Allowed Transitions and Selection Rules

According to Equations (5.14) and (5.15), we see that the probability of a particular transition depends on the electric dipole matrix element μ, given by Equation (5.12). These transitions, which are induced by interactions of the electric dipole element with the electric field of the incident radiation, are called *electric dipole transitions*. Therefore, electric dipole transitions are allowed when $\mu \neq 0$.

Let us now examine the circumstances in which $\mu \neq 0$. In expression (5.12), the operator $\mathbf{r}$ has odd parity (i.e., $\mathbf{r} = -(-\mathbf{r})$). Because of this, the matrix element μ (or $\langle \psi_f|\mathbf{r}|\psi_i\rangle$) is zero whenever the wavefunctions ψ_i and ψ_f have the same parity. In fact, the integral in Equation (5.12) can be written as a sum of two contributions at points $\mathbf{r}$ and $-\mathbf{r}$. For wavefunctions with the same parity, these contributions are equal but opposite, so that $\mu = 0$. Consequently, electric dipole transitions are allowed when the initial and final states have opposite parity but they are forbidden for states with equal parity. Remembering that the parity of a state is given by $(-1)^l$, l being the orbital quantum number, this is the well-known *Laporte selection rule* that we know from quantum mechanics. In any case, as we will note below, optical centers in some crystals do not rigorously obey this law.

Provided that a transition is forbidden by an electric dipole process, it is still possible to observe absorption or emission bands induced by a *magnetic dipole transition*. In this case, the transition proceeds because of the interaction of the center with the magnetic field of the incident radiation. The interaction Hamiltonian is now written as $H = \mathbf{u}_m \cdot \mathbf{B}$, where $\mathbf{u}_m$ is the magnetic dipole moment and $\mathbf{B}$ is the magnetic field of the radiation.

EXAMPLE 5.2   *Electric dipole versus magnetic dipole transitions.*

In this example, we will roughly estimate the order of magnitude for the intensity ratio of the electric dipole to magnetic dipole transitions. Of course, we will assume that both processes are allowed (which, as shown below, is not possible for a given transition) and that the same excitation intensity is used.

The electric dipole transition probability (expression (5.10)) can be roughly approximated by

$$(P_{if})_e \propto (E_0 \times p)^2 \approx (E_0 \times ea)^2$$

where the magnitude of the electric dipole moment of the valence electron, p, has been approximated by the product of the electronic charge, e, and the radius of the atom, a. E_0 is the amplitude of the electric field due to the incident radiation.

In a similar way, we can approximate the transition probability for a magnetic dipole allowed transition by

$$(P_{if})_m \propto (B_0 \times u_m)^2 \approx (B_0 \times \beta)^2$$

where B_0 is the amplitude of the magnetic field due to the incident radiation and the magnetic moment of the valence electron is approximated by the Bohr magneton, $\beta = 9.27 \times 10^{-24}$ A m^2. This approach is reasonable, as $\mathbf{u}_m = -g\beta\mathbf{J}$, g being the giromagnetic factor (1/2 for a free electron) and $\mathbf{J} = \mathbf{L} + \mathbf{S}$ is the total angular momentum. Thus, the Bohr magneton clearly defines the value of $\mathbf{u}_m$, so that $|\boldsymbol{\mu}_m| \approx \beta$.

Taking into account that, for a plane wave, $E_0 = B_0 \times c$ (where c is the speed of light) and assuming $a \approx 0.5$ Å (the Bohr radius), the ratio between the electric and magnetic dipole probabilities is given by

$$\frac{(P_{if})_e}{(P_{if})_m} \approx \frac{(B_0 c \times ea)^2}{(B_0 \times \beta)^2} = \left(\frac{cea}{\beta}\right)^2 \approx 10^5$$

We see that the electric dipole allowed transitions are, in general, much more intense than the magnetic dipole allowed transitions. In fact, the magnetic dipole contribution to an optical transition of a center dominated by an electric dipole character is usually completely masked by the much more intense electric dipole transitions.

As shown in Example 5.2, magnetic dipole transitions are much weaker than electric dipole transitions. Nevertheless, when a radiative transition is forbidden by an electric dipole process, it may happen due to a magnetic dipole process. In fact, the magnetic dipole moment is a function with even parity[2] and so magnetic dipole transitions are allowed between states with the same parity, while they are forbidden between states of different parity. Consequently, a forbidden electric dipole transition is allowed by a magnetic dipole process and vice versa. For instance, this implies that the 3d → 3d transition of the Ti^{3+} ion, treated in the previous section, is forbidden at the electric dipole order, while it is permitted by a magnetic dipole process, as both states have the same parity ($l = 2$). In any case, this rule is only rigorously true for specific centers (i.e., for ions in specific crystal environments) where l is still a 'good quantum number.'

Up to this print, we have simply reexamined the selection rules between electronic configurations (s, p, d, ...) of free ions. It is also instructive to recall the rules for other interaction orders, such as electron–electron and spin–orbit interactions:

- For ^{2S+1}L terms (i.e., states where the total spin S and the total orbital angular momentum L are good quantum numbers), the allowed transition are $\Delta S = 0$ and $\Delta L \neq 0$.

- For $^{2S+1}L_J$ states (i.e., states where $J = L + S$ is a good quantum number), the selection rule is $\Delta J = \pm 1, 0$, but $J = 0 \rightarrow J = 0$ is forbidden.

5.3.3 Polarized Transitions

We will now consider an absorption (or emission) process in which the incident (or emitted) light is linearly polarized; that is, with its electric vector **E** oscillating along one defined direction. Let us suppose that this direction is parallel to one symmetry axis, say the x-axis, of our center. Assuming an electric dipole allowed transition, the interaction Hamiltonian between the light and the center is now expressed as $H = \mathbf{p} \cdot \mathbf{E} = p_x E$. Hence the matrix element (5.12) must now be written as $\langle \psi_f | ex | \psi_i \rangle$. Because of the nature of the electronic states, it can happen that, for instance, the matrix element $\langle \psi_f | ex | \psi_i \rangle$ is zero while the matrix element $\langle \psi_f | ey | \psi_i \rangle$ is nonzero. This would mean that the **E** ∥ **x** light would not be absorbed (emitted) in an optical transition between the states $|\psi_i\rangle$ and $|\psi_f\rangle$, while the **E** ∥ **y** polarized light would be absorbed (emitted). In fact, the overall absorption (or emission) probability P_{if} given in Equation (5.14) must involve the summation over all allowed polarizations:

$$P_{if} = (P_{if})_x + (P_{if})_y + (P_{if})_z \tag{5.16}$$

where $(P_{if})_x$, $(P_{if})_y$, and $(P_{if})_z$ are the absorption (or emission) probabilities with the electric field along the x-, y-, and z-axes of the center, respectively. The evaluation of these probabilities for centers in crystals is, in general, a complicated task. However,

---

[2] This can be roughly seen by considering the classical view of the valence electron, where this electron describes a circular orbit of radius r around the nucleus. In this case, the magnitude of the magnetic dipole moment is proportional to the area of the circular orbit and $|\mathbf{u}_m| \propto r^2$, so that $\mathbf{u}_m(\mathbf{r}) = \mathbf{u}_m(-\mathbf{r})$.

as we will see in Chapter 7, the allowance of a given transition in a specific polarization can be obtained by symmetry considerations of the center.

5.3.4 The Probability of Spontaneous Emission

Formula (5.14) gives the absorption probability, P_{if}, for our simple two energy level system. By Equation (5.15), we know that this is equal to the probability of stimulated emission, P_{fi}. However, as we have shown in Section 1.4, we know that once the system has been excited it can also return spontaneously to the ground state by emitting a photon with an energy corresponding to the energy separation between the two energy levels. The probability per second for this spontaneous decay, or the radiative rate A, has already been defined in Equation (1.17). This probability can be estimated by perturbation theory and also by means of an elegant thermodynamic argument, due to Einstein. In the latter approach (developed in Appendix A3), it is assumed that our two-level system is introduced into a blackbody radiation box, whose walls are kept at a fixed temperature T. Then the probability of spontaneous emission is related to the absorption probability through the Einstein coefficients. For an electric dipole process, the probability of spontaneous emission is given by

$$A = \frac{n\omega_0^3}{3\pi\hbar\varepsilon_0 c_0^3} |\boldsymbol{\mu}|^2 \tag{5.17}$$

where ω_0 corresponds to the transition frequency of the system.

By Equation (5.17), we see that the probability of spontaneous emission is proportional to $|\boldsymbol{\mu}|^2$, so we can use the same selection rules as previously established for absorption and stimulated emission to predict the allowance of spontaneous emission. Moreover, it can be noted that A is proportional to ω_0^3 and so, for a small energy separation between the two levels of our system, the radiative emission rate A should also be small. In this case, nonradiative processes, described by A_{nr} (see Equation (1.17)), can be dominant so that no emitted light is observed.

EXAMPLE 5.3   *Radiative lifetimes for electric dipole and magnetic dipole transitions.*

The order of magnitude of the probability of spontaneous emission in the visible range can be estimated from Equation (5.17). We consider a typical dielectric medium with $n = 1.5$ at a wavelength in the middle of the visible range, $\lambda_0 = 500$ nm ($\omega_0 = 3.8 \times 10^{15}$ s^{-1}). If we make the approximation $|\boldsymbol{\mu}| = ea$, where $e = 1.6 \times 10^{-19}$ C is the electronic charge and $a \approx 1$ Å is the atomic radius, then Equation (5.17) gives an electric dipole probability of spontaneous emission of

$$(A)_e = 9 \times 10^7 \text{ s}^{-1} \approx 10^8 \text{ s}^{-1}$$

which corresponds to a radiative lifetime of $(\tau_0)_e = 1/(A)_e \approx 10^{-8} \text{ s}^{-1} = 10 \text{ ns}$.

As shown in Example 5.2, it is easy to obtain that $(A)_e/(A)_m \approx 10^5$, where $(A)_m$ is the probability of spontaneous emission for a magnetic dipole transition. Thus, using the previous estimation of $(A)_e$, we obtain that, for a magnetic dipole transition,

$$(A)_m \approx 10^{-5}(A)_e = 10^3 \ \text{s}^{-1}$$

which corresponds to a radiative lifetime of $(\tau_0)_m \approx 10^{-3} \ \text{s}^{-1} = 1$ ms.

Recall that the radiative lifetime, $\tau_0 = 1/A$, can be determined from Equation (1.20) by measuring the fluorescence lifetime τ from a luminescence decay-time experiment, and provided that the nonradiative rate A_{nr} is known. For processes where the nonradiative rate is negligible ($A_{nr} \approx 0$), $\tau \cong \tau_0$ and so we will measure lifetimes in the range of *nanoseconds* for electric dipole transitions and lifetimes in the range of *microseconds* for magnetic dipole transitions.

5.3.5 The Effect of the Crystal on the Transition Probabilities

At this point, it should be mentioned that the absorption and emission probabilities given by expressions (5.14) and (5.17) were derived for a center in a diluted medium. As we mentioned in Chapter 4 (Section 4.3), in dense media such as crystals, a suitable correction must be introduced to take into account the actual local electric field $\mathbf{E}_{loc}$ acting on the valence electrons of our absorbing center due to the electromagnetic incoming wave. This electric field may be different from the average electric field in the medium $\mathbf{E}_0$ that we considered in Equation (5.11). To take this effect into account, we must replace the factor $|\boldsymbol{\mu}|^2$ by $(\mathbf{E}_{loc}/\mathbf{E}_0)^2 |\boldsymbol{\mu}|^2$ in the transition probabilities given by Equation (5.14) and (5.17). After taking this effect into account, the spontaneous emission probability formula (5.17) can be written in a convenient way as

$$A = \frac{1}{4\pi\varepsilon_0} \frac{4n\omega_0^3}{3\hbar c_0^3} \left(\frac{\mathbf{E}_{loc}}{\mathbf{E}_0}\right)^2 |\boldsymbol{\mu}|^2 \qquad (5.18)$$

A second effect that must be considered for centers in crystals is that the eigenfunctions ψ_i and ψ_f of the initial (ground) and final (excited) states needed to evaluate the matrix element $\boldsymbol{\mu}$ are no longer those of the free ion but those of the ion in the crystal. Therefore, the selection rules established above must be modified. For instance, the Laporte rule can be strongly affected, as l is no longer a good quantum number.

Fortunately, the selection rules can be well established by group theory considerations, as will be shown in Section 7.5. Nevertheless, we can say that, in general, the Laporte rule is still fulfilled in local environments with inversion symmetry, as for the AB_6 center of Figure 5.1. This is because in these symmetries the eigenfunctions still preserve the parity character of the free-ion eigenfunctions. However, in centers with noninversion symmetry some crystal field Hamiltonian terms produce admixing of states belonging to different electronic configurations; the eigenfunctions do not have a well-defined parity and the Laporte selection rule is no longer valid.

For a better understanding of this effect, it is illustrative to come back to the case of Ti^{3+} (a $3d^1$ outer electronic configuration) in Al_2O_3, discussed in Section 5.2. The optical absorption band at around 500 nm (see Figure 5.4), due to the $d(t_{2g})$ → $d(e_g)$ intraconfigurational transition, should be forbidden in a regular octahedral environment with inversion symmetry. However, the actual arrangement of the O^{2-} ligand ions around the Ti^{3+} ions in Al_2O_3 does not correspond to the regular octahedral arrangement of Figure 5.1. Rather, the oxygen ions are displaced in respect to the ligand positions of the AB_6 center displayed in Figure 5.1, producing a local environment of trigonal symmetry. This is a noninversion symmetry. As a result, the $d(t_{2g})$ → $d(e_g)$ absorption is allowed at the electric dipole order, as the t_{2g} and e_g states correspond to an admixing of the 3d electronic configuration with the higher electronic configurations 4s, 4p, and so on: this releases us from the Laporte parity selection rule. Such types of transitions are called *electric dipole forced transitions*.

5.3.6 Oscillator Strength: Smakula's Formula

Let us now establish a way in which to relate $|\mu|^2$, or the transition probability given in Equation (5.14), with experimental measurements, such as the absorption spectrum.

Considering our single two energy level center, it is easy to understand that the area under the absorption spectrum, $\int \alpha(\omega) \, d\omega$, must be proportional to both $|\mu|^2$ and the density of absorbing centers, N. In order to build up this proportionality relationship, it is very common to use a dimensionless quantity, called the *oscillator strength, f*. This magnitude has already been introduced in the previous chapter (Section 4.3), when treating the classical Lorentz oscillator. It is defined as follows:[3]

$$f = \frac{2m\omega_0}{3\hbar e^2} \times |\mu|^2 \qquad (5.19)$$

where m is the electronic mass and ω_0 is the frequency at the absorption peak. Classically, the oscillator strength f represents the number of electric dipole oscillators that can be stimulated by the radiation field (in the dielectric dipole approximation) and has a value close to one for strongly allowed transitions. Comparing expressions (5.18) and (5.19), we can see how the oscillator strength f is directly correlated with the spontaneous emission probability A by

$$A = \frac{1}{4\pi\varepsilon_0} \frac{2\omega_0^2 e^2}{mc_0^3} \left[\left(\frac{\mathbf{E}_{loc}}{\mathbf{E}}\right)^2 n \right] \times f \qquad (5.20)$$

---

[3] This expression is for singly degenerate energy levels. For an initial state with degeneracy g, this expression must be multiplied by $1/g$.

Now, it can be demonstrated (see Appendix A4) that the area under the absorption spectrum is related to f and to the density of absorbing centers, N, by

$$\int \alpha(\omega)\,d\omega = \frac{1}{4\pi\varepsilon_0}\frac{2\pi^2 e^2}{mc_0}\left[\left(\frac{\mathbf{E}_{loc}}{\mathbf{E}}\right)^2\frac{1}{n}\right] \times f \times N \qquad (5.21)$$

For ions in crystals of high symmetry, as in the case of our reference octahedral AB_6 center, the correction factor is $\mathbf{E}_{loc}/\mathbf{E}_0 = (n^2 + 2)/3$ (Fox, 2001), where n is the refractive index of the medium. Although this correction factor is not strictly valid for centers of low symmetry, it is often used even for these centers. Thus, assuming this local field correction and inserting numerical values for the different physical constants, expression (5.21) becomes

$$N(\text{cm}^{-3})f = 54.1\frac{n}{(n^2+2)^2}\int \alpha(\omega)(\text{cm}^{-1})\,d\omega \qquad (5.22)$$

which is known as *Smakula's formula* for electric dipole absorption processes. This relation is very useful in experimentally determining the oscillator strength (or $|\boldsymbol{\mu}|^2$) from the absorption spectrum of a given system, if the density of absorbing centers is known. Smakula's formula can also be used to determine the density of absorbing centers from the absorption spectrum if the oscillator strength is known, as in the next example.

EXAMPLE 5.4   *Figure 5.8 shows the absorption spectrum of a NaCl crystal containing color centers generated by irradiation. The band peaking at 443 nm is related to the so-called F centers, for which the oscillator strength is $f = 0.6$. From this absorption band, determine the density of the F centers that have been produced by the irradiation process. Assume a refractive index of $n = 1.6$ for NaCl.*

The F center is an electron trapped at a negative Cl^- vacant site. These centers can be created in NaCl by irradiation or by additive coloration, as shown in the next chapter (Section 6.5). The band at 443 nm corresponds to a certain concentration, N, of F centers that have been introduced by the irradiation process. The other band peaking at about 280 nm is related to other types of color center (which are formed by F center aggregation), beyond the scope of this example.

From the full width at half maximum (FWHM), $\Delta E = 0.34$ eV, of the F band, we can determine the FWHM in frequency units, $\Delta\omega$:

$$\Delta\omega = \frac{\Delta E}{\hbar} = \frac{0.34\text{ eV}}{6.59 \times 10^{-16}\text{ eVs}} = 5.5 \times 10^{14}\text{ s}^{-1}$$

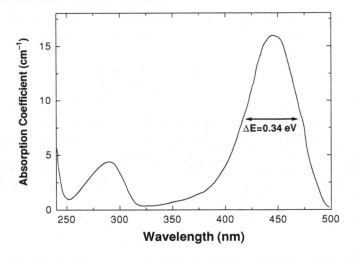

Figure 5.8   The absorption spectrum at 77 K of an irradiated NaCl crystal. The absorption band at 443 nm is due to the F centers generated after irradiation. The full width at half maximum of this band is indicated on the figure.

We can now make a rough approximation for the area under the F-band absorption spectrum, by $\int \alpha(\omega) \, d\omega \approx \alpha_{max} \times \Delta\omega$, where $\alpha_{max} = 16.1 \text{ cm}^{-1}$ is the absorption coefficient at the peak.

Considering an electric dipole absorption process, which is in agreement with the oscillator strength $f = 0.6$, we can use Equation (5.22) to estimate N, the concentration of absorbing F centers:

$$N \approx 54.1 \frac{n}{(n^2 + 2)^2} \alpha_{max} \times \Delta\omega \times \frac{1}{f} = 5.75 \times 10^{16} \text{ cm}^{-3}$$

Thus, a concentration of about 6×10^6 F centers per cm^{-3} has been produced by a certain radiation dose.

5.4 DYNAMIC INTERACTION: THE CONFIGURATIONAL COORDINATE DIAGRAM

In the previous sections, we have considered that the optical center is embedded in a static lattice. In our reference model center AB_6 (see Figure 5.1), this means that the A and B ions are fixed at equilibrium positions. However, in a real crystal, our center is part of a vibrating lattice and so the environment of A is not static but dynamic. Moreover, the A ion can participate in the possible collective modes of lattice vibrations.

In order to understand the dynamic effects on optical spectra, we have to consider that the ion A is *coupled* to the vibrating lattice. This means that the neighboring B ions can vibrate about some average positions and this affects the electronic states of the ion A. Additionally, the environment can also be affected by changes in the electronic state of the ion A. For example, when the ion A changes its electronic state, the ligand B ions may adopt new average positions and the nature of their vibrations about these new average positions may not be the same as for the initial electronic state.

To take the above-mentioned ion–lattice coupling into account, the full ion-plus-lattice system must be considered, so that the static Hamiltonian given by Equation (5.2) must be replaced by

$$H = H_{FI} + H_{CF} + H_L \tag{5.23}$$

where H_L is the Hamiltonian describing the lattice (the kinetic and potential energies of the lattice) and $H_{CF} \equiv H_{CF}(\mathbf{r}_i, \mathbf{R}_l)$ is the crystalline field Hamiltonian, which now depends on both $\mathbf{r}_i$ (the coordinates of the valence electrons of ion A) and $\mathbf{R}_l$ (the coordinates of the B ions). Consequently, at variance with the static case, the crystalline field term *couples* the electronic and ionic motions. In fact, the eigenfunctions are now functions of the electronic and ionic coordinates, $\psi \equiv \psi(\mathbf{r}_i, \mathbf{R}_l)$ ($l = 1, 2, \ldots 6$ for the AB_6 center).

The solution of the Schrödinger equation, $H\psi = E\psi$, is now much more complicated and some approximations must be considered to take into account different coupling strengths.

For *weak coupling* between the ion A and the lattice, the crystalline field is very weak ($H_{CF} \approx 0$) and so the electronic and ionic motions are practically independent of each other. In this case, weak side bands are sometimes observed in addition to those corresponding to pure electronic transitions. These additional bands are due to the participation of ion A in the vibrational motion of the lattice, which leads to Doppler-modulated absorption or emission bands. A good example of such a kind of coupling is given in Figure 5.9, which shows a single absorption line (a transition between two defined electronic energy levels) of Yb^{3+} ions in $LiNbO_3$ accompanied by a series of phonon side bands. This side-band spectrum is essentially coincident with the Raman spectrum of lithium niobate.

For *strong coupling*, the band shape is strongly affected, as occurs for the emission band of Cr^{3+} ions in $LiNbO_3$ shown in Figure 5.9. This emission band, which corresponds to a single transition in the static lattice, appears strongly broadened as a result of the phonon coupling. To account for such a type of coupling, we have to invoke the *configurational coordinate model*. This model was briefly introduced in Section 1.4 (Figure 1.10), to explain the Stokes shift between absorption and emission bands. The configurational coordinate model is based on two main approximations:

(i) The first approximation, due to Born and Oppenheimer (1927), is called the *adiabatic approximation*. It considers that the ions move very slowly in comparison to the valence electrons, so that the electronic motion takes place at a given

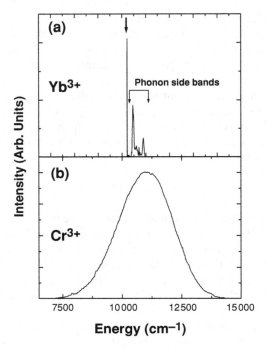

Figure 5.9 Two examples of dynamic induced band-shape effects. (a) Weak coupling; an absorption single line of the Yb^{3+} ion in $LiNbO_3$ (denoted by an arrow) is accompanied by the appearance of phonon side bands (reproduced with permission from Montoya *et al.*, 2001) (b) Strong coupling: the broadband luminescence of the Cr^{3+} ion in $LiNbO_3$ (reproduced with permission from Camarillo *et al.*, 1992).

nuclear coordinate (electrons move without perceiving changes in the nuclear positions). This approximation is reasonable as nuclei are much heavier than electrons, and therefore move on a much slower timescale. Within the adiabatic approximation, the nuclear and electronic motions can be solved independently and so the electronic energy can be drawn as a function of the distances A–B. In other words, this means that the eigenfunctions can be factored as follows:

$$\psi = f(\mathbf{r}_i, \mathbf{R}_l) \cdot \chi(\mathbf{R}_l) \qquad (5.24)$$

where the $f(\mathbf{r}_i, \mathbf{R}_l)$ are the electronic functions for the static case (at the coordinates $\mathbf{R}_l$) and the $\chi(\mathbf{R}_l)$ are the vibrational wavefunctions with regard to the motion of the ions.

(ii) The second approximation is just to limit our attention to only one representative (ideal) mode of vibration instead of the many possible modes. It is usual to choose the so-called *breathing mode*, in which the ligand B ions pulsate radially 'in and out' about the A central ion. In this case, we need only one nuclear coordinate, called the *configurational coordinate Q*, which corresponds to the distance A–B. However, we know that there is a large number of vibrational modes in a crystal. Thus, in general, the configurational coordinate can represent the average amplitude of one of these modes or perhaps a linear combination of several of them.

In any case, under this assumption, the eigenfunctions given by Equation (5.24) are simplified as follows:

$$\psi = f(\mathbf{r}_i, Q) \cdot \chi(Q) \tag{5.25}$$

The solution of the Schrödinger equation of our one-coordinate dynamic center (Henderson and Imbusch, 1989) leads to potential energy curves for the ground i (initial) and excited f (final) states as diagrammatically represented in Figure 5.10. Such a diagram is called a *configurational coordinate diagram*. The curves in the diagram represent the interionic interaction potential energy (the Morse potential), while the horizontal lines over each curve represent the set of permitted discrete energies (phonon states). It must be noted that the equilibrium position coordinates, Q_0 and Q'_0, are different for the ground and excited states; and also that, at distances close to the equilibrium coordinate, the interionic potential curves can be approximated by parabolas (broken curves) according to the harmonic oscillator approximation.

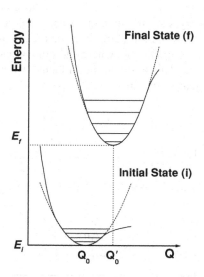

Figure 5.10 The configurational coordinate diagram for the AB_6 center oscillating as a breathing mode. The broken curves are parabolas within the approximation of the harmonic oscillator. The horizontal full lines are phonon states.

In this approximation, the B ions pulsate in harmonic oscillation around the equilibrium positions and, consequently, the interionic potential energies, $E_i(Q)$ and $E_f(Q)$, of the ground and excited states are given by

$$E_i(Q) = E_i + \tfrac{1}{2}M\Omega_i^2 (Q - Q_0)^2 \tag{5.26}$$

$$E_f(Q) = E_f + \tfrac{1}{2}M\Omega_f^2 (Q - Q_0')^2 \tag{5.27}$$

where M is an effective oscillating mass and Ω_i and Ω_f are characteristic vibrational frequencies for the ground and excited states, respectively. These frequencies are considered to be different, as the center can pulsate at different frequencies in the ground and excited states.

The discrete energy levels sketched as horizontal lines on each potential curve of Figure 5.10 are consistent with the quantized energy levels (phonon levels) of a harmonic oscillator. For each harmonic oscillator at frequency Ω, the permitted phonon energies are given by

$$E_n = \left(n + \tfrac{1}{2}\right)\hbar\Omega \tag{5.28}$$

where $n = 0, 1, 2, \ldots$ and so on. Each one of these states is described by a harmonic oscillator function $\chi_n(Q)$, and the probability of finding an electron at Q in the nth vibrational state is given by $|\chi_n(Q)|^2$. As a relevant example, in Figure 5.11 the shapes of $|\chi_n(Q)|^2$ for the states $n = 0$ and $n = 20$ have been drawn. This reveals that for the lowest energy state the maximum amplitude probability occurs at the equilibrium position, Q_0, while for a large n value the maximum probability occurs at the configurational coordinates Q where the corresponding vibrational energy crosses with the parabola. This has a strong influence when determining the shape functions of the spectra, as shown in the next section.

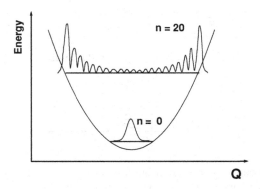

Figure 5.11   The shape of the $|\chi_0(Q)|^2$ and $|\chi_{20}(Q)|^2$ functions for a quantum harmonic oscillator.

5.5 BAND SHAPE: THE HUANG–RHYS COUPLING PARAMETER

Let us now deal with the shape of the optical (absorption and emission) bands as a result of the previously discussed strong ion–lattice coupling. For this purpose, we consider a simplified two electronic energy states center, in which the initial i and final f states are both described by harmonic oscillators of the same frequency Ω and with minima at Q_0 and Q_0' respectively, as shown in Figure 5.12.

In the spirit of the adiabatic approximation, the transitions between two vibrational states (belonging to initial and final electronic states) must occur so rapidly that there is no change in the configurational coordinate Q. This is known as the *Frank Condon principle* and it implies that the transitions between i and f states can be represented by vertical arrows, as shown in Figure 5.12. Let us now assume our system to be at absolute zero temperature (0 K), so that only the phonon level $n = 0$ is populated and all the absorption transitions depart from this phonon ground level to different phonon levels $m = 0, 1, 2, \ldots$ of the excited state. Taking into account Equation (5.25), the absorption probability from the $n = 0$ state to an m state varies as follows:

$$P_{if}(n = 0 \rightarrow m) \propto |\langle f(Q)|H_{\text{int}}|i(Q)\rangle|^2 \times |\langle \chi_m(Q)|\chi_0(Q)\rangle|^2 \qquad (5.29)$$

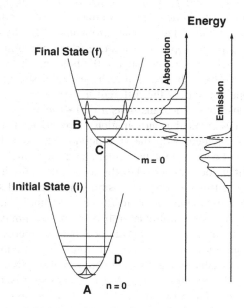

Figure 5.12 A configurational coordinate diagram with which to analyze transitions between two electronic states. Harmonic oscillators at the same frequency Ω are assumed for both states. The absorption and emission band profiles are sketched based on the $0 \rightarrow m$ (absorption) and $n \leftarrow 0$ (emission) relative transition probabilities (see the text). For simplicity, the minima of these parabolas, Q_0 and Q_0', are not represented.

where H_{int} is the interaction Hamiltonian between the light and valence electrons of ion A, and $i(Q)$ and $f(Q)$ are the electronic functions of the ground and excited states, respectively. If we assume that these electronic functions do not vary significantly compared to their values at Q_0, we can write the previous expression as follows:

$$P_{if}(n = 0 \rightarrow m) \propto |\langle f(Q_0) | H_{int} | i(Q_0) \rangle|^2 \times |\langle \chi_m(Q) | \chi_0(Q) \rangle|^2 \qquad (5.30)$$

where the term $\langle f(Q_0) | H_{int} | i(Q_0) \rangle$ would correspond to the matrix element in the static case (rigid lattice), and the term $|\langle \chi_m(Q) | \chi_0(Q) \rangle|^2$ gives the relative absorption probability due to the overlap between the $\chi_0(Q)$ and $\chi_m(Q)$ vibrational functions.

The overall absorption probability P_{if} at 0 K involves a summation of the probabilities from the ground vibrational state, $n = 0$, to all of the excited m states ($m = 0, 1, 2, \ldots$), and so it can be written as follows:

$$P_{if} \propto \sum_m |\langle f(Q_0) | H_{int} | i(Q_0) \rangle|^2 \times |\langle \chi_m(Q) | \chi_0(Q) \rangle|^2 = |\langle f(Q_0) | H_{int} | i(Q_0) \rangle|^2$$

$$(5.31)$$

as the $\chi_m(Q)$ functions form an orthonormal set and we can verify that $\sum_m |\langle \chi_m(Q) | \chi_0(Q) \rangle|^2 = 1$. Therefore, according this simple model, the absorption probability between the i and f electronic states is equal to that given by expression (5.14) for the case of a static lattice. The only effect of the dynamic lattice is to change the band shape, but not the full absorption probability.[4] Let us now investigate this dynamically induced band shape, which is due to the different overlapping factors between the vibrational m states and the $n = 0$ vibrational state, $|\langle \chi_m(Q) | \chi_0(Q) \rangle|^2$.

In Figure 5.12, we have displayed the shape of the absorption (and emission) band as an envelope curve over the different $n = 0 \rightarrow m = 0, 1, 2, \ldots (m = 0 \rightarrow n = 0, 1, 2, \ldots)$ transitions. The transitions $n = 0 \leftrightarrow m = 0$ are called *zero-phonon lines*, as they occur without the participation of phonons. Thus, the zero-phonon absorption line is coincident with the zero-phonon emission line. The maximum in the absorption band occurs at the particular energy for which there is a maximum overlap factor, and it has been indicated by the arrow AB in Figure 5.12. It corresponds to a transition from A (the equilibrium position in the ground state, $n = 0$), where the amplitude probability is maximum, to B (a cross-point in an excited level of the terminal state, f), where the amplitude probability is also a maximum. By the same reasoning, the maximum emission intensity occurs at an energy that corresponds to the arrow CD in Figure 5.12. As shown in this figure, the maximum in the emission peak occurs at a lower energy than the maximum in absorption, thus explaining the *Stokes shift*, already defined in section 1.4. The Stokes shift is an important feature, as it avoids a strong overlap between the absorption and emission bands. Otherwise, the emitted light would be reabsorbed by the emitting center.

---

[4] This is only true for vibration modes that preserve inversion symmetry but not for vibronic transitions (see p. 000).

The Stokes shift is usually measured in terms of the lateral displacement of the ground and excited state parabolas, $\Delta Q = Q_0' - Q_0$ (see Figure 5.10); a large Stokes shift between the ground and excited states indicates a strong difference in the electron–host coupling for these two electronic states. To somehow quantify the difference in the electron–lattice coupling, a dimensionless parameter, S, called the *Huang–Rhys parameter*, is defined as follows:

$$\tfrac{1}{2}M\Omega^2(\Delta Q)^2 = S\hbar\Omega \tag{5.32}$$

Thus, the Huang–Rhys parameter is a measure of the Stokes shift (or the displacement between the ground and excited parabolas). In fact, from Figure 5.12 it can be shown that

$$E_a - E_e = 2\tfrac{1}{2}M\Omega^2(\Delta Q)^2 - 2\tfrac{1}{2}\hbar\Omega = (2S - 1)\hbar\Omega \tag{5.33}$$

where $E_a - E_e$ is the Stokes shift energy, E_a being the energy at the absorption maximum (corresponding to AB in Figure 5.12) and E_e being the energy at the emission maximum (corresponding to CD in Figure 5.12).[5]

The absorption (emission) band shape at 0 K can be estimated from Equation (5.30) if the square of the overlap integral of harmonic functions, $|\langle \chi_m(Q)|\chi_0(Q)\rangle|^2$, is known for each excited (terminal) m level. Using the wavefunctions of the harmonic oscillator, these overlap functions can be expressed as a function of S:

$$|\langle \chi_m(Q)|\chi_0(Q)\rangle|^2 = e^{-S} \times \frac{S^m}{m!} \tag{5.34}$$

Thus, we can predict the relative intensity of each $0 \rightarrow m$ absorption line using this expression together with Equation (5.30):

$$I_{0 \rightarrow m} = e^{-S} \times \frac{S^m}{m!} \tag{5.35}$$

We must recall that that, according to Equation (5.31), the full absorption intensity (the area under the absorption band) is independent of S. Thus, the factor e^{-S} (which corresponds to $I_{0\rightarrow 0}$) represents the fraction of the absorption intensity taken by the zero-phonon line, while the intensity $I_{0\rightarrow 1} = e^{-S} \times S$ represents the fractional intensity related to the $0 \rightarrow 1$ transition, $I_{0\rightarrow 2} = e^{-S} \times (S^2/2)$ the fractional intensity of the $0 \rightarrow 2$ transition, and so on.

At this point, we are able to predict the low-temperature optical (absorption and emission) band shapes for different *coupling* strengths (i.e., different Huang–Rhys parameters), as shown in the next example.

---

[5] Equation (5.33) is not correct for weak coupling. For this case, $S \cong 0$ and then $E_a = E_e$, which is at variance with this expression. Expression (5.35) can be considered valid only for $(2S - 1) \geq 0$; that is, $E_a - E_e > 0$.

EXAMPLE 5.5   *Sketch the absorption and emission spectra at 0 K for bands with zero-phonon line at 600 nm, a coupling with an unique breathing mode of energy 200 cm^{-1} and a Huang–Rhys parameter of S = 1.*

A zero-phonon line ($n = 0 \rightarrow m = 0$ transition) at 600 nm corresponds to a wavenumber of 16 666 cm^{-1}. Then, the $n = 0 \rightarrow m$ transitions will occur at $(16\,666 + m200)$ cm^{-1}. Using Equation (5.35), we can calculate the relative intensity for each $0 \rightarrow$ m transition. The intensities and wavenumbers calculated for the different transitions are listed in Table 5.1. They have been used to display the absorption spectrum at 0 K (see Figure 5.13).

Table 5.1   The relative intensities and wavenumbers of the different $0 \rightarrow 1, 2, 3, 4, 5$ (absorption and emission) transitions for the spectroscopic parameters given in Example 5.5.

Transition	Absorption (cm^{-1})	Emission (cm^{-1})	Intensity ($\times$ 1/e)
$0 \rightarrow 0$	16 666	16 666	1
$0 \rightarrow 1$	16 866	16 466	1
$0 \rightarrow 2$	17 066	16 266	0.5
$0 \rightarrow 3$	17 266	16 066	0.17
$0 \rightarrow 4$	17 466	15 866	0.04
$0 \rightarrow 5$	17 666	15 666	0.008

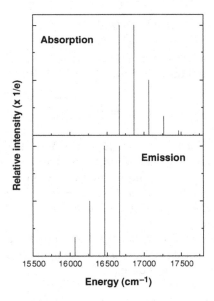

Figure 5.13   Simulated absorption and emission spectra (at 0 K) for bands with zero-phonon line at 16 666 cm^{-1} (600 nm), $S = 1$, and coupling with a phonon of 200 cm^{-1}.

The 0 K emission spectrum occurs from the lowest excited state vibrational level ($m = 0$) to levels $n = 0, 1, 2, \ldots$ Thus the relative intensities are the same as those calculated for the absorption spectrum, but now the lines appear on the low-energy side of the zero-phonon line. The $n \leftarrow 0$ emission line appears at $(16\,666 - n200)$ cm^{-1}. Thus, we can calculate all of the data needed to display the emission spectrum. These data are included in Table 5.1 and the emission spectrum is displayed in Figure 5.13, together with the absorption spectrum. It can be noted that the emission spectrum appears to be Stokes shifted to lower energies by an energy of 200 cm^{-1}, in agreement with Equation (5.33) (we have considered the maximum absorption energy at the average absorption wavenumber, 16 766 cm^{-1}, and the maximum emission energy at the average emission wavenumber, 16 566 cm^{-1}).

Following the same steps as in Example 5.5, in Figure 5.14 we have displayed the band shape expected (at 0 K) for $S = 0$ (a weak coupling case) and $S = 7$ (a strong coupling case), together with the one discussed in the previous example for $S = 1$, for comparison. We can see that within this model (one single breathing mode frequency), the case $S = 0$ only consists of a zero-phonon line, as the $0 \rightarrow 0$ transition takes up all of the band intensity, $I_{0 \rightarrow 0} = 1$. Thus, this spectrum corresponds to a transition between pure electronic states. As S increases, the relative intensity in the zero-phonon line decreases and this is accompanied by the appearance of vibrational side bands,

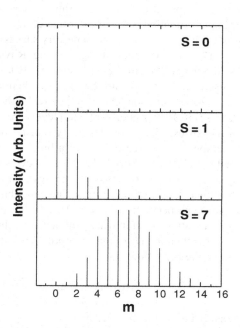

Figure 5.14 The low-temperature band shape, plotted as intensity versus m (final state), expected for different coupling (Huang–Rhys) parameters.

which are observed at energies $m\hbar\omega$ above the zero-phonon line. For high S values, the band becomes broad and structureless.

In the majority of cases, broad optical bands are indicative of strong ion–lattice coupling, while sharp optical bands are indicative of weak ion–lattice coupling. However, it is important to say that for a given center that shows strong electron–lattice coupling, there can be bands with $S \gg 0$ and bands with $S \approx 0$. This, for instance, is the case for Cr^{3+} ions in crystals, which are discussed in the next chapter (Section 6.4), where the ion strongly interacts with the host crystal in the ground state, $^4A_2$, and in the excited states. However, the ion–lattice interaction of the Cr^{3+} ion in the excited state ^2E is almost identical to that of the ground state, so that the $^4A_2 \rightarrow {}^2E$ absorption band is very narrow (see, for instance, the absorption spectrum of Al_2O_3:Cr^{3+} in next chapter; Figure 6.9) because of a weak *difference in the ion–lattice coupling* between the ground state and the excited state ($S \approx 0$). On the other hand, in other excited states, such as the $^4T_2$, the Cr^{3+} ion shows a large difference in ion–lattice coupling in respect to the ground state $^4A_2$, so that $^4A_2 \rightarrow {}^4T_2$ transitions are broad (a large S parameter).

In our previous approach using the configurational coordinate model, we have assumed only a single vibrational frequency (mode). However, in reality, we know that there is a wide spectrum of vibrational frequencies rather than a single breathing mode frequency. Thus, we can expect that in practice the shapes of low-temperature spectra will be somewhat different to that predicted by our single-frequency model. To take this spread of phonons into account and thus to have a better approach to actual spectra, a good approximation is to consider that each $0 \rightarrow m$ transition has a linewidth equal to $m\hbar\Omega$. This approach has become very successful in explaining the $^4A_2 \rightarrow {}^4T_2$ broad absorption band of ruby, using a Huang–Rhys parameter of $S = 7$ and a breathing mode of energy $250\ cm^{-1}$ (see Exercise 5.10).

The band shapes discussed up to now are for spectra taken at absolute zero temperature, so that only the lowest phonon level is populated in the departure state. For higher temperatures, $T > 0$ K, higher-energy phonon levels are populated, at the cost of a depopulation of the lowest vibrational level. By reinspecting Figure 5.12, it is easy to see that any temperature increase leads to a broader absorption (emission) band, since the excited $n = 1, 2, 3, \dots$ ($m = 1, 2, 3, \dots$) levels are populated and so they also participate in the absorption (emission) process. Now, a thermal average over the excited vibrational levels of the initial state must be carried out, so that the absorption (emission) can occur from $n = 1, 2, 3, \dots$ ($m = 1, 2, 3, \dots$) phonon states. Taking this thermalization effect into account, it can be shown (Henderson and Imbusch, 1989) that the bandwidth ΔE of the absorption/emission bands varies with temperature according to the following:

$$\Delta E(T) \approx \Delta E(0)\sqrt{\coth(\hbar\Omega/2kT)} \qquad (5.36)$$

where $\Delta E(0)$ is the bandwidth at 0 K and $\hbar\Omega$ is the energy of the coupling phonon.

Equation (5.36) shows that the bandwidth increases with temperature. However, within this simple model, the center of gravity and the full intensity (the area under the band) are expected to remain constant, as does the total population.

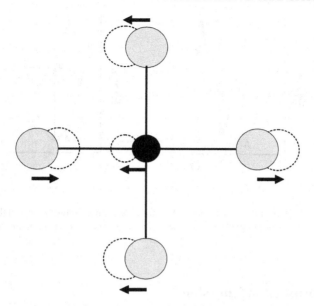

Figure 5.15 A two-dimensional center, showing a mode vibration destroying the inversion symmetry of the static lattice. The black circle represents the central ion (A), while the other circles represent the ligand (B) ions.

We showed in Equation (5.31) that the area under a given band is independent of the ion–lattice coupling and, as said in the previous paragraph, it remains constant with increasing temperature. However, most broadband transitions of transition metal ions in solids are of electric dipole nature, caused by dynamic lattice distortions of odd symmetry. These types of transitions are called *vibronic transitions*, and they occur in cases where the dynamic symmetry breaks the inversion symmetry of the center in the static lattice. They are, in fact, dynamically forced electric dipole transitions. Figure 5.15 shows, as an illustrative example, how an odd-parity mode of vibration of a two-dimensional AB_4 center destroys its inversion symmetry in the static lattice. This dynamic distortion can lead to vibronic transitions. Obviously, the intensity of these transitions (the area under the optical band) is dependent on the strength of the coupling, and it is also influenced by temperature changes.

5.6 NONRADIATIVE TRANSITIONS

Once a center has been excited we know that, in addition to luminescence, there is the possibility of nonradiative de-excitation; that is, a process in which the center can reach its ground state by a mechanism other than the emission of photons. We will now discuss the main processes that compete with direct radiative de-excitation from an excited energy level.

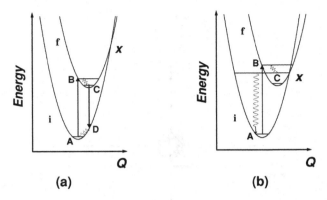

Figure 5.16 Configurational coordinate diagrams to explain (a) radiative and (b) nonradiative (multiphonon emission) de-excitation process. The sinusoidal arrows indicate the nonradiative pathways.

5.6.1 Multiphonon Emission

The most important nonradiative de-excitation process that competes with radiative de-excitation is that due to *multiphonon emission*. We can make use of the configurational coordinate diagram to qualitatively explain this multiphonon de-excitation process. In Figure 5.16 two configurational coordinate diagrams, corresponding to cases of strong electron–lattice coupling, have been displayed. In both cases, the crossover point, X, between the parabolas of the initial (i) and final (f) states is indicated.

In Figure 5.16(a), the maximum of the absorption spectrum (at 0 K) corresponds to the line AB, the maximum overlap of the vibrational wavefunctions. This transition terminates in the vibrational level corresponding to point B, which is below the crossover point, X. This process is followed by a fast down-relaxation by multiphonon emission to the point C, from which the emission originates. Thus, the emission spectrum has its maximum at an energy corresponding to the line CD. Finally, another multiphonon emission process takes place by down-relaxation from D to the departing point A.

In Figure 5.16(b), which corresponds to a higher Huang–Rhys parameter, the crossover point X is at lower energy than point B. Thus, the center is down-relaxed by multiphonon emission to the vibrational state corresponding to the crossover point X. This level is degenerate in energy, as it belongs to both the ground and the excited state parabolas. From this vibrational level, the de-excitation probability is much larger through the phonon states of parabola i than through the phonon states of parabola f. Therefore, as the vibrational level corresponding to point C is not populated, luminescence does not take place. The system returns to the ground state (point A) by means of a full nonradiative multiphonon relaxation through parabola i.

The configurational coordinate model also provides a qualitative explanation for the extinction of the luminescence when the temperature is raised, a process commonly known as *thermal quenching*. This process occurs as a result of the thermal population of vibrational levels higher than those corresponding to points A (in the ground state parabola) and B (in the excited state parabola). The population of these levels means that the level at X (in Figure 5.16(a)) can also be populated. The system then returns to the ground state via nonradiative decay, giving rise to a thermally quenched luminescence.

In cases of weak coupling ($S \approx 0$), there is no crossover point between the ground and excited state parabolas (assuming the same shape for both of them), so that the multiphonon emission nonradiative mechanism cannot be explained by invoking the configurational coordinate model. This is the case for trivalent rare earth ions in crystals, which will be treated in next chapter, in Section 6.3.

5.6.2 Energy Transfer

An excited center can also relax to the ground state by *nonradiative energy transfer* to a second nearby center. The sequence of such an energy transfer process has been sketched in Figure 5.17: (1) The D center, called the *donor*, absorbs the excitation light $h\nu_D$ and shifts to an excited state D^*: (2) and (3) This donor center then relaxes to its ground state by transmitting its excitation energy to a second center A, called the *acceptor*, which shifts to an excited state A^*: (4) Finally, this acceptor center relaxes to its ground state by emitting its own characteristic radiation $h\nu_A$. It is important to remark that in the transfer process (2) $\rightarrow$ (3) no photons are emitted by the D donor ion. Such a process, sometimes called *internal luminescence* or *radiative energy transfer*, is not of interest for practical applications.

Nonradiative energy transfer is very often used in practical applications, such as to enhance the efficiency of phosphors and lasers. A nice example is the commercial phosphor $Ca_5(PO_4)_3$ (FCl), which is doubly activated by Sb^{3+} and Mn^{2+} ions. When the phosphor is singly activated by Mn^{2+} ions, it turns out to be very inefficient, due to the weak absorption bands of the divalent manganese ion. However, coactivation with Sb^{3+} ions produces a very intense emission from the Mn^{2+} ions, because the Sb^{3+} ions (the donor centers) efficiently absorb the ultraviolet emission (253.6 nm) of

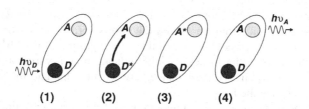

Figure 5.17   Sequential steps for a nonradiative energy transfer process (see the text).

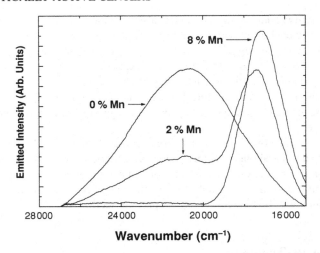

Figure 5.18   The emission spectra of the phosphor $Ca_5(PO_4)_3$ (FCl) activated with a fixed concentration of Sb^{3+} (0.01 mol/mol Ca) and coactivated with two different concentrations of Mn^{2+} (0.02 and 0.08 mol/mol Ca) (reproduced with permission from Nakazawa, 1998).

the Hg atoms inside the fluorescent tube (see Figure 2.3 in Chapter 2) and transfer part of this energy to the Mn^{2+} ions (the acceptor centers), which emit their characteristic fluorescence.

Figure 5.18 shows clear evidence of $Sb^{3+} \rightarrow Mn^{2+}$ nonradiative energy transfer. In this figure, three emission spectra of the lamp phosphor $Ca_5(PO_4)_3$ (FCl):Sb:Mn are shown under excitation with light lying within the Sb^{3+} absorption bands; one spectrum for a singly doped crystal with a given concentration of Sb^{3+} and two spectra for doubly doped crystals with the same concentration of Sb^{3+} and two different concentrations of Mn^{2+}. For the singly doped crystal (0 % Mn), the typical blue emission (peaking at around $21\,000\,cm^{-1}$) of Sb^{3+} ions is observed. This emission decreases when codoping with a certain concentration (2 %) of Mn^{2+} and the appearance of a new yellow–orange band (peaking at around $17\,000\,cm^{-1}$) characteristic of Mn^{2+} ions is observed. By a further increase in the Mn^{2+} content (8 %), the blue emission of Sb^{3+} disappears completely and only the typical fluorescence of Mn^{2+} ions is observed. This indicates that, for this Mn^{2+} concentration, all of the Sb^{3+} excited ions transfer their energy to Mn^{2+} ions. However, for the lower Mn^{2+} concentration (2 %), both the blue emission (from Sb^{3+}) and the yellow–orange emission (from Mn^{2+}) are simultaneously observed, because in this case a portion of the Sb^{3+} excited ions relaxes by photon emission. These results indicate how a variation in the Sb/Mn content permits us to change the balance between the blue and the yellow–orange emissions, and hence the lamp spectrum.

To allow energy transfer, some interaction mechanism between the excited donor D^* and the acceptor A is needed. In fact, the probability of energy transfer from the donor centers to the acceptor centers can be written as follows (Föster, 1948; Dexter, 1953):

$$P_t = \frac{2\pi}{\hbar} |\langle \psi_D \psi_{A*} | H_{\text{int}} | \psi_{D*} \psi_A \rangle|^2 \int g_D(E) g_A(E) \, dE \qquad (5.37)$$

where ψ_D and ψ_{D*} denote the wavefunctions of the donor center in the ground and excited states, respectively, ψ_A and ψ_{A*} are the wavefunctions of the acceptor center in the ground and excited states, respectively, and H_{int} is the D–A interaction Hamiltonian. The integral in Equation (5.37) represents the overlap between the normalized donor emission line-shape function, $g_D(E)$, and the normalized acceptor absorption line-shape function, $g_A(E)$. Indeed, this term is needed for energy conservation, being a maximum when D and A are centers with coincident energy levels, a case that is called *resonant energy transfer* (see Figure 5.19(a)).

However, when D and A are different centers, it is usual to find an energy mismatch between the transitions of the donor and acceptor ions (see Figure 5.19(b)). In this case, the energy transfer process needs to be assisted by lattice phonons of appropriate energy, $\hbar\Omega$, and this is usually called *phonon-assisted energy transfer*. In these energy transfer processes, electron–phonon coupling must be also taken into account, together with the interaction mechanism responsible for the transfer.

The interaction Hamiltonian that appears in Equation (5.37) can involve different types of interactions; namely, *multipolar* (electric and/or magnetic) *interactions* and/or a quantum mechanical *exchange interaction*. The dominant interaction is strongly dependent on the separation between the donor and acceptor ions and on the nature of their wavefunctions.

For *electric multipolar interactions*, the energy transfer mechanism can be classified into several types, according to the character of the involved transitions of the donor (D) and acceptor (A) centers. Electric dipole–dipole (d–d) interactions occur when the transitions in D and A are both of electric dipole character. These processes correspond, in general, to the longest range order and the transfer probability varies with $1/R^6$, where R is the separation between D and A. Other electric multipolar interactions are only relevant at shorter distances: dipole–quadrupole (d–q) interaction varies as $1/R^8$, while quadrupole–quadrupole interaction varies as $1/R^{10}$.

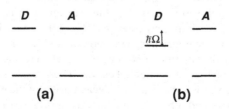

Figure 5.19 The energy-level schemes of donor D and acceptor A centers for (a) resonant energy transfer and (b) phonon-assisted energy transfer.

The dependence on R of the transfer probability due to electric multipolar interactions can be written in a general way as follows:

$$P_t(R) = \frac{\alpha_{dd}}{R^6} + \frac{\alpha_{dq}}{R^8} + \frac{\alpha_{qq}}{R^{10}} + \dots \tag{5.38}$$

where the factors α_{dd}, α_{dq}, and α_{qq} weighting the different interactions depend on the different spectroscopic magnitudes of the D and A centers, including the overlap factor given in Equation (5.37) (Henderson and Imbusch, 1989). If electric dipole transitions are allowed for both the D and the A centers, it happens that $\alpha_{dd} > \alpha_{dq} > \alpha_{qq}$, and so the dipole–dipole interaction is dominant. However, if the electric dipole transitions are not completely allowed either for the D or for the A centers, it is probable that the higher order interaction processes, d–q or q–q, will have larger transfer probabilities at short distances because of the higher order exponent of R.

Energy transfer probabilities due to *multipolar magnetic interactions* also behave in a similar way to that previously discussed for multipolar electric interactions. Thus, the transfer probability for a magnetic dipole–dipole interaction also varies with $1/R^6$, and higher order magnetic interactions are only influential at short distances. In any case, the multipolar magnetic interactions are always much less important than the electric ones.

Exchange interactions only occur if the donor and acceptor ions are close enough for direct overlap of their electronic wavefunctions. Consequently, energy transfer due to quantum mechanical exchange interactions between the D and A ions is only important at very short distances (nearest neighbor positions). In fact, the transfer probability varies similarly to the overlap of the wavefunctions: $P_t \propto e^{-2R/L}$, where L is an average of the radii of the D^* and A ions ($L \approx 10^{-10}$ m).

Whatever the particular mechanism of energy transfer is, the fluorescence lifetime of the donor center, τ_D, is affected as a result of any energy transfer process to an acceptor. Therefore, expression (1.20) must be rewritten for the lifetime of the donor ions as follows:

$$\frac{1}{\tau_D} = \frac{1}{(\tau_D)_0} + A_{\mathrm{nr}} + P_t \tag{5.39}$$

where $(\tau_D)_0$ is the radiative lifetime of the donor ion, A_{nr} is the nonradiative rate due to multiphonon relaxation, and P_t is the transfer rate due to energy transfer. In most real cases, the emission decay intensity $I(t)$ of the donor ion is not exponential. This is due to the statistical distribution of both D and A ions in real systems, which produces a statistical distribution of D–A distances. As a result, the transfer rate is not homogeneous, leading to nonexponential curves. These aspects are clearly manifested in Figure 5.20, which shows the temporal decay curves of the emission intensity, $I(t)$, for both Nd^{3+} (donor) and Yb^{3+} (acceptor) ions in doubly (Yb, Nd) doped yttrium aluminum borate ($YAl_3(BO_3)_4$) laser crystals. The purpose of this coactivation was

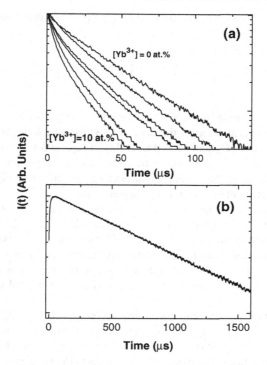

Figure 5.20 The time evolution of the fluorescence of Nd^{3+} (donor) and Yb^{3+} ions (acceptor) in the $YAl_3 (BO_3)_4$ crystal. (a) The decay time of Nd^{3+} for different Yb^{3+} ion concentrations (for simplicity only the lowest and highest Yb^{3+} concentrations are indicated). (b) The time evolution of the fluorescence of Yb^{3+} ion (5 at.%) for a 5 at.% concentration of Nd^{3+} ions (unpublished data).

to increase the laser efficiency of the Yb^{3+} ion in this crystal by pumping into the Nd^{3+} absorption bands. As a result of Nd → Yb energy transfer, the $I(t)$ decay curves of the Nd^{3+} donor ions are not exponential.

By inspection of Figure 5.20(a), it can also be seen how the fluorescence signal decays faster when the concentration of the acceptor (Yb^{3+}) is increased. In fact, increasing the Yb^{3+} concentration reduces the average Nd^{3+}–Yb^{3+} distance, which increases the transfer rate.

The shape of the $I(t)$ curves of the donor centers carries very useful information about the nature of the interaction process. Assuming that the acceptors A are randomly distributed at various distances from the donor centers D, the Japanese scientists Inokuti and Hirayama (1965) investigated the shape of the donor decay-time curves $(I(t))$ for the different multipolar interactions and also for the exchange interaction.

For electric multipolar interactions, the shape of $I(t)$ is given by:

$$I(t) = I(0) \, \exp\left[-\frac{t}{\tau_D} - \Gamma\left(1 - \frac{3}{s}\right)\frac{C}{C_0}\left(\frac{t}{\tau_D}\right)^{3/s}\right], \qquad s = 6, 8, 10 \qquad (5.40)$$

where $\Gamma(\ )$ is the gamma function, C is the concentration of the acceptor A centers, C_0 is a critical concentration of the acceptor (A) for which the transfer probability, P_t, equals the donor (D) emission probability, $1/\tau_D$, and s is a factor that is 6 for d–d energy transfer, 8 for d–q energy transfer, and 10 for q–q energy transfer. Thus, by fitting the shape of the experimental decay-time curves to Equation (5.40), the dominant interaction mechanism can be determined.

The decay-time curve of the acceptor centers (A) under excitation in the donor centers (D) is also nonexponential. This is evident in Figure 5.20(b), where the time evolution of the emission intensity of the Yb^{3+} ions (the acceptors) in $YAl_3(BO_3)_4$ is shown under excitation with light absorbed by the Nd^{3+} ions (the donors). The Yb^{3+} decay-time curve shows an initial rise time, due to the excitation via energy transfer from the Nd^{3+} ions, followed by the characteristic exponential decay of the Yb^{3+} ions.

Of course, energy transfer between donor centers $(D–D)$ and/or acceptor centers $(A–A)$ can also take place. Energy transfer between centers of the same type is very often called *energy migration,* as the excitation energy can migrate through several ions. If energy migration is considered, in addition to the $D–A$ energy transfer process, the complexity of the problem increases considerably as a consequence of sequential energy transfer processes.

5.6.3 The Concentration Quenching of Luminescence

In principle, an increase in the concentration of a luminescent center in a given material should be accompanied by an increase in the emitted light intensity, this being due to the corresponding increase in the absorption efficiency (see expression (1.15)). However, such behavior only occurs up to a certain critical concentration of the luminescent centers. Above this concentration, the luminescence intensity starts to decrease. This process is known as *concentration quenching* of luminescence.

Figure 5.21 shows a manifestation of this effect for the main infrared luminescence (around 1.5 μm) of Er^{3+} ions in CaF_2 layer crystals (crystals grown by molecular beam epitaxy) (Daran *et al.*, 1994). In this figure, the infrared emission intensity of Er^{3+} ions in CaF_2 layers is shown as a function of the Er content for a fixed excitation intensity. Up to about 35 % Er concentration, the emitted intensity grows monotonously with the Er content. At higher erbium doping levels the

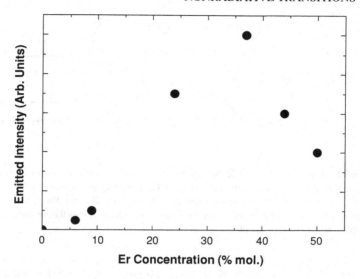

Figure 5.21 The integrated emission intensity of Er^{3+} ions (in the infrared range, 1.5–1.6 μm) as a function of Er concentration (reproduced with premission from Daran *et al.*, 1994).

luminescence decreases, so that above a 50 % Er concentration the emission cannot be detected.

In general, the origin of luminescence concentration quenching lies in a very efficient energy transfer among the luminescent centers. The quenching starts to occur at a certain concentration, for which there is a sufficient reduction in the average distance between these luminescent centers to favor energy transfer. Two mechanisms are generally invoked to explain the luminescence concentration quenching:

(i) Due to very efficient energy transfer, the excitation energy can migrate about a large number of centers before being emitted. However, even for the purest crystals, there is always a certain concentration of defects or trace ions that can act as acceptors, so that the excitation energy can finally be transferred to them. These centers can relax to their ground state by multiphonon emission or by infrared emission. Thus, they act as an energy sink within the transfer chain and so the luminescence becomes quenched, as illustrated in Figure 5.22(a). These kinds of centers are called *killers* or *quenching traps*.

(ii) Concentration quenching can also be produced without actual migration of the excitation energy among the luminescent centers. This occurs when the excitation energy is lost from the emitting state via a *cross relaxation* mechanism. This kind of relaxation mechanism occurs by resonant energy transfer between two identical adjacent centers, due to the particular energy-level structure of these

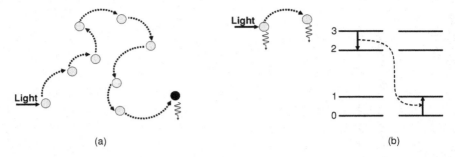

Figure 5.22 Schemes of possible mechanisms for luminescence concentration quenching: (a) energy migration of the excitation along a chain of donors (circles) and a killer (black circle), acting as nonradiative sink; (b) cross relaxation (including an illustrative energy-level diagram) between pairs of centers. (Sinusoidal arrows indicate nonradiative decay or radiative decay from another excited level.)

centers. Figure 5.22(b) shows a simple possible energy-level scheme involving cross-relaxation. We suppose that for isolated centers radiative emission ($3 \rightarrow 0$) from level 3 dominates. However, for two nearby similar centers a resonant energy transfer mechanism can occur in which one of the centers (the one acting as donor) transfers part of its excitation energy – for instance, $E_3 - E_2$ – to the other center (the one acting as acceptor). This resonant transfer becomes possible due to the particular disposition of the energy levels, in which the energy for the transition $3 \rightarrow 2$ is equal to that for the transition $0 \rightarrow 1$. As a result of the cross-relaxation, the donor center will be in the excited state 2, while the acceptor center will reach the excited state 1. Then, from these states a nonradiative relaxation or emission of photons with energies other than $E_3 - E_0$ will occur; in any case, the $3 \rightarrow 0$ emission will be quenched.

As the concentration quenching results from energy transfer processes, the decay time of the emitting ions is reduced when one concentration quenching mechanism occurs. In general, this decay-time reduction is much easier to measure than the reduction in the quantum efficiency. In fact, the easiest way to detect luminescence concentration quenching is to analyze the lifetime of the excited centers as a function of the concentration. The critical concentration is that for which the lifetime starts to be reduced.

Finally, it is important to mention that besides the possibility of energy transfer, a high concentration of centers can lead to new kinds of centers, such as clusters formed by aggregation or coagulation of individual centers. Thus, these new centers can have a different level scheme to that of the isolated centers, giving rise to new absorption and emission bands. This is, of course, another indirect mechanism of concentration quenching for the luminescence of the isolated centers, as happens in the next example.

EXAMPLE 5.6   *The low-temperature emission lifetime of a very low concentration of Cr^{3+} ions in ruby (Al_2O_3 :Cr) is about 4 ms. For a 1% Cr_2O_3 concentrated sample, the measured lifetime becomes 1.7 ms. Estimate the loss of efficiency due to this concentration quenching of the luminescence.*

For simplicity, we assume a quantum efficiency of $\eta = 1$ for the low-concentration emission at low temperature. Thus, the measured lifetime is just the radiative lifetime; $\tau_0 = 4$ ms. For high Cr concentrations, we can write the following general expression:

$$\frac{1}{\tau} = \frac{1}{\tau_0} + P_t$$

where P_t represents the average probability of transfer to the killer. In this case, the killers are pairs of Cr^{3+} ions, called exchange coupled pairs. The lifetime of the 1 % concentrated crystal is reduced to $\tau = 1.7$ ms (the decay-time curve is nearly exponential) because of the nonvanishing term P_t due to the energy transfer from Cr^{3+} ions to Cr^{3+}–Cr^{3+} pairs. Therefore, the new efficiency is

$$\eta = \frac{1/\tau_0}{1/\tau} = \frac{1.7\,\text{ms}}{4\,\text{ms}} = 0.425$$

Consequently, the efficiency is reduced by a factor of 57.5 % with regard to the quantum efficiency of the diluted system.

5.7 ADVANCED TOPIC: THE DETERMINATION OF QUANTUM EFFICIENCIES

As we have seen in this chapter, the luminescence quantum efficiency of an emitting system is an important magnitude, as it defines the ratio between the direct radiative de-excitation rate from a given level and the total de-excitation rate (see Equation (1.21)). This total rate includes the other possible de-excitation rates, such as multiphonon relaxation, energy transfer, and concentration quenching. In principle, according to the definition given in Equation (1.13), the quantum efficiency from an emitting level can be experimentally obtained from a luminescence experiment by measuring the number of photons that need to be absorbed per emitted photon; that is, by measuring the absorbed and emitted intensities at the same time. However, due to the scattered nature of the emitted radiation and the difficulty in determining the exact excited volume of sample, luminescence experiments are not commonly used for determining quantum efficiencies. On the other hand, the analysis of the energy delivered by multiphonon relaxation is very useful in determining the quantum efficiency of diluted systems (i.e., systems for which the concentration of centers is low enough for energy transfer and migration processes to be neglected). In such

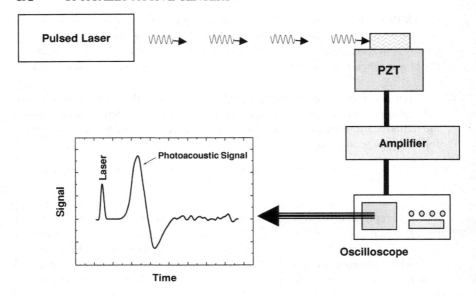

Figure 5.23 The scheme of an experimental arrangement for PA spectroscopy. The inset shows a typical PA signal generated as a result of illuminating an absorbing sample with pulsed light.

cases, there are only two methods of de-excitation: luminescence and multiphonon relaxation. The later process leads to heat generation and then the heating rate can be followed by various thermal sensing techniques (calorimetric techniques, thermal refractive-index gradients, surface deformation, etc.) (Tam, 1986).

A very common heating sensing technique used in condensed matter is *photo-acoustic* (PA) *spectroscopy*, which is based on detection of the acoustic waves that are generated after a pulse of light is absorbed by a luminescent system. These acoustic waves are produced in the whole solid sample and in the coupling medium adjacent to the sample as a result of the heat delivered by multiphonon relaxation processes.

Figure 5.23 shows a typical experimental arrangement for PA measurements. A pulsed laser is used as an excitation source and the PA signal is collected by an acoustic detector, such as a microphone or a resonant piezoelectric transducer (PZT), located as close as possible to the sample. The later detector has the advantage that it can be glued to the solid sample, permitting good acoustic transmission from the solid to the detector. The PA signal is then conveniently amplified and recorded by a digital oscilloscope. The acoustic wave consists of a series of compressions and expansions that produce an oscillating PA signal with the typical shape displayed in Figure 5.23. This signal is delayed with respect to the laser excitation pulse because of the time that the acoustic wave needs to reach the acoustic detector.

Let us now briefly discuss the method of PA spectroscopy for measuring quantum efficiencies, considering the simple two-level system of Figure 5.24(a).

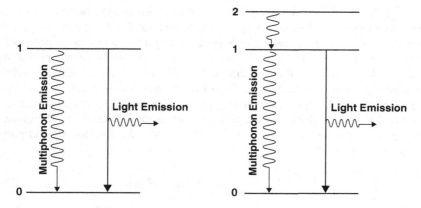

Figure 5.24   (a) The energy-level scheme of a two-level system, showing the two possible de-excitation process (luminescence and multiphonon relaxation). (b) A three-level scheme, needed to calibrate the acoustic signal and then to determine the quantum efficiency.

If, after the excitation pulse is absorbed, the excited state 1 reaches a population density N_1, then the heat released per unit volume H is given by

$$H = N_1 E_{10}(1 - \eta) \tag{5.41}$$

where $E_{10} = E_1 - E_0$ is the energy separation between the excited and ground states and η is the luminescence quantum efficiency. The heat released gives rise to an acoustic wave signal, whose intensity S is proportional to the heating rate generated, so that we can write:

$$S = K N_1 E_{10}(1 - \eta) \tag{5.42}$$

where K is a proportionality constant. This constant depends on the volume of the excited sample, on the coupling of the sample with the acoustic detector through the air, and on the photoacoustic response of the system.

Therefore, the quantum efficiency η can be determined from Equation (5.42), provided that S is measured and both K and N_1 have previously been determined. N_1 can be determined in a simple way if the absorption coefficient of the Sample, α_{10}, at the energy E_{10} is measured:

$$N_1 \cong I_0 \tau_p \alpha_{10} \tag{5.43}$$

where I_0 is the excitation intensity (measured in photons per second per unit area) and τ_p is the excitation pulse duration. However, the constant K in Equation (5.42) is difficult to calculate and it must be calibrated by means of another independent measurement.

A method devoted to avoiding the evaluation of this proportionality constant has been established for the case of optical centers in solids (Rodriguez *et al.*, 1993), provided that the excited state 1 can be populated by a full nonradiative process from an upper level (level 2 in Figure 5.24(b)). The method consists of the measurement of the photoacoustic signal at two different excitation wavelengths corresponding to the photon energies E_{10} and $E_{20} = E_2 - E_0$. Excitation with photons of energy E_{20} populates state 2 and this is followed by a fast nonradiative down-relaxation to the emitting level 1, for which the quantum efficiency is to be determined. Therefore, the PA signal S_2 after excitation with a pulse of intensity I_0, duration τ_p, and photon-energy E_{20} is given by

$$S_2 = K\alpha_{20}I_0\tau_p(E_{20} - \eta E_{10}) \tag{5.44}$$

The PA signal, S_1, obtained under excitation with a pulse of the same intensity and duration but of photon energy E_{10}, is easily found by using expressions (5.42) and (5.43). The ratio between these two photoacoustic signals, S_1/S_2, is given by:

$$\frac{S_1}{S_2} = \frac{K I_0\tau_p\alpha_{10}(1 - \eta)E_{10}}{K I_0\tau_p\alpha_{20}(E_{20} - \eta E_{10})} \tag{5.45}$$

After simplification, this expression becomes:

$$\frac{S_1}{S_2} = \frac{\alpha_{10}(1 - \eta)E_{10}}{\alpha_{20}(E_{20} - \eta E_{10})} \tag{5.46}$$

Thus, once the absorption coefficients α_{10} and α_{20} and the ratio between the PA signals S_1/S_2 have been measured, the quantum efficiency η can be determined from the previous Equation (5.46).[6]

Of course, this model can only be applied if there is an upper level (level 2) that populates the emitting level 1 by a full nonradiative process. This model was initially applied (Rodriguez *et al.*, 1993) to determine the luminescence quantum efficiency of Eu^{2+} ions in potassium chloride crystals. Eu^{2+} ions show two broad absorption bands in crystals attributed to transitions from the $4f^7$ ground electronic configuration to the $4f^65d^1$ excited electronic configuration. The octahedral crystal field of the Cl^- ligand ions on the Eu^{2+} ion splits its $5d^1$ electronic configuration into two components, t_{2g} and e_g (see Section 5.2). The emission of Eu^{2+} ions in KCl is related to the transition from the lowest-energy excited component (t_{2g}) to the ground state (the $4f^7$ configuration). Therefore, this transition corresponds to the $1 \rightarrow 0$ transition in Figure 5.24(b). Excitation into the higher component (e_g or level 2 in Figure 5.24(b)) also leads to the $1(t_{2g}) \rightarrow 0(4f^7)$ emission spectrum because of a full nonradiative relaxation from the $2(e_g)$ level to the $1(t_{2g})$ level. Consequently, the

---

[6] Of course, in real cases this simple method must be adequate to take into account the heat released due to the band shape and the Stokes shift.

previously described model can be applied to this system. A quantum efficiency of $\eta = 1$ was determined for the emission of Eu^{2+} in KCl.

In general, it is possible to find such a full nonradiative level for other ions in crystals, so that the model can be applied.

EXERCISES

5.1. Consider the absorption spectrum of Ti^{3+} ions in $Al_2 O_3$ (sapphire), shown in Figure 5.4. (a) From this spectrum, estimate the amount $10Dq$. (b) Assuming that the $3d^1$ valence electron describes a circular trajectory of 0.1 nm radius and that there is an octahedral environment of point charges (the O^{2-} ions), estimate the Ti^{3+}–O^{2-} distance in sapphire.

5.2. Suppose that a high hydrostatic pressure is applied to a crystal sample of sapphire ($Al_2 O_3:Ti^{3+}$). How do you expect that the absorption spectrum of Ti^{3+} (Figure 5.4) will be affected?

5.3. The Ti^{3+} ion shows a broad absorption band that peaks at around 560 nm in phosphate glasses. This band is associated with the $t_{2g} \rightarrow e_g$ transition of these ions in an octahedral environment of O^{2-} ions. (a) Estimate the crystalline field splitting, $10Dq$. (b) Suppose that a new glass is prepared, in which the Ti^{3+} ions are located in a cubic environment of O^{2-} ions (TiO_8 centers) but maintaining the same Ti–O distance as for the phosphate glass. Determine the crystalline field splitting $10Dq$ for this new glass and the peak wavelength due to the $e_g \rightarrow t_{2g}$ absorption of Ti^{3+} ions.

5.4. The cross section, σ, for a given transition, which was defined in Equation (1.6), can be redefined (see Appendix A4) as $\sigma = P/I_p$, where P is the transition rate (or transition probability) and $I_p = I/\hbar\omega$ is the photon flux of the incident wave. Using expressions (5.14) and (5.17), demonstrate that $\sigma = (\lambda/2)^2 g(\omega)/\tau_0$, where λ is the wavelength (in the medium) of an electromagnetic wave whose frequency corresponds to the center of the transition, τ_0 is the radiative lifetime, and $g(\omega)$ is the line-shape function.

5.5. Cr^{3+} ions in aluminum oxide (the ruby laser) show a sharp emission (the so-called R_1 emission line) at 694.3 nm. To a good approximation, the shape of this emission is Lorentzian, with $\Delta\nu = 330$ GHz at room temperature. (a) Provided that the measured peak transition cross section is $\sigma = 2.5 \times 10^{-20}$ cm^2 and the refractive index is $n = 1.76$, use the formula demonstrated in the previous exercise to estimate the radiative lifetime. (b) Since the measured room temperature fluorescence lifetime is 3 ms, determine the quantum efficiency for this laser material.

5.6. The A–B distance, a, in an AB_6 center embedded in a host lattice shows a temperature variation given by $a(T) = a(0) \times (1 + \alpha T)$, where α is the thermal dilatation coefficient of the lattice and $a(0)$ is the distance A–B at absolute zero temperature. If the ion A has an outer d^1 electronic configuration, explain how a temperature increase will affect the absorption spectrum peak of such a center.

5.7. Ni^{2+} ions show a broad absorption blue band that peaks at 405 nm in MgF_2 crystal, a tunable laser system. For a concentration of 2×10^{20} cm^{-3} Ni^{2+} ions, the absorption coefficient measured at 405 nm is 7.2 cm^{-1} and the full width at half maximum is 8.2×10^{13} Hz. Taking into account a refractive index of 1.39 for the MgF_2 crystal, estimate the oscillator strength, f, of the transition responsible for the mentioned absorption band.

5.8. A certain transition metal ion presents two optical absorption bands in a host crystal whose zero-phonon lines are at 600 nm and 700 nm, respectively. The former band has a Huang–Rhys parameter $S = 4$, while for the latter $S = 0$. Assuming coupling with a phonon of 300 cm^{-1} for the two bands: (a) display the 0 K absorption spectrum (absorption versus wavelength) for such a transition metal ion; (b) display the emission spectra that you expect to obtain under excitation in both absorption bands; and (c) explain how you expect these two bands to be affected by a temperature increase.

5.9. A host material is activated with a certain concentration of Ti^{3+} ions. The Huang–Rhys parameter for the absorption band of these ions is $S = 3$ and the electronic levels couple with phonons of 150 cm^{-1}. (a) If the zero-phonon line is at 522 nm, display the 0 K absorption spectrum (optical density versus wavelength) for a sample with an optical density of 0.3 at this wavelength. (b) If this sample is illuminated with the 514 nm line of a 1 mW Ar^+ CW laser, estimate the laser power after the beam has crossed the sample. (c) Determine the peak wavelength of the 0 K emission spectrum. (d) If the quantum efficiency is 0.8, determine the power emitted as spontaneous emission.

5.10. The $^4A_{2g} \rightarrow {}^4T_{2g}$ transition of the Cr^{3+} ion in aluminum oxide (ruby) leads to an important broad absorption band for pumping the ruby laser. The shape of this absorption band (at 0 K) can be described by a strong ion–lattice coupling case with a Huang–Rhys parameter of $S = 7$ (a band shape shown in Figure 5.14). Draw the shape of this absorption band (intensity versus m), but taking into consideration that each side-band feature at energy $m\hbar\Omega$ above the zero-phonon line has a full width $m\hbar\Omega$. Then compare it with the band shape given in Figure 5.14. The shape of this spectrum fits the actual shape of the $^4A_{2g} \rightarrow {}^4T_{2g}$ absorption band of ruby much better (see Figure 6.9 in Chapter 6).

5.11. The fluorescence lifetime of Eu^{2+} in fluorite (CaF_2) crystals is 700 ns. When these crystals are codoped with Sm^{2+} ions, the lifetime of the Eu^{2+} ions is reduced to 150 ns. Determine the energy transfer rate from the Eu^{2+} ions to the Sm^{2+} ions.

5.12. The fluorescence lifetime, τ, of Nd^{3+} in $Ca_3Ga_2Ge_3O_{12}$ crystal (a solid state laser) is measured as a function of the neodymium concentration, C_{Nd}, giving

Table E5.12   The lifetimes of Nd^{3+} ions in
$Ca_3Ga_2Ge_3O_{12}$ for different Nd concentrations

C_{Nd} (at. %)	τ (μs)
0.1	240
1	240
2	161
8	39
16	16

the results shown in Table E5.12. Assuming a quantum efficiency of $\eta = 1$ for the most diluted system (0.1 % Nd): (a) determine the radiative lifetime; (b) determine the quantum efficiency for all the other concentrations; (c) make a plot of η versus C_{Nd} and explain the observed evolution; and (d) explain the Nd concentration that you would choose to make a laser system.

REFERENCES AND FURTHER READING

Ballhausen, C. J., and Gray, H. B., *Molecular Orbital Theory*, W. A. Benjamin, Inc., New York (1965).

Born, M., and Oppenheimer, J. R., *Ann. Phys.* (Liepzig), **84**, 457 (1927).

Camarillo, E., Tocho, J., Vergara, I., Dieguez, E., García Solé, J., and Jaque, F., *Phys. Rev. B*, **45**(9), 4600 (1992).

Daran, E., Legros, R., Muñoz-Yagüe, A., and Bausá, L. E., *J. Appl. Phys.*, **76**(1), 270 (1994).

Dexter, D. L., *J. Chem. Phys.*, **21**, 836 (1953).

Föster, T., *Ann. Phys.* (Leipzig), **2**, 55 (1948).

Fox, M., *Optical Properties of Solids*, Oxford University Press, Oxford (2001).

Henderson, B., and Imbusch, G. F., *Optical Spectroscopy of Inorganic Solids*, Clarendon Press, Oxford (1989).

Inokuti, M., and Hirayama, F., *J. Chem. Phys.*, **43**, 1978 (1965).

Montoya, E., Agulló-Rueda, F., Manotas, S., García Solé, J., and Bausá, L. E., *J. Lumin.*, **94–95**, 701 (2001).

Nakarawa, E., in *Phosphor Handbook*, eds. S. Shionoya and W. M. Yen, CRC Press (1998).

Rodriguez, E., Tocho, J. O., and Cussó, F., *Phys., Rev. B*, **47**(21), 14 049 (1993).

Svelto, O., *Principles of Lasers*, Plenum Press, New York (1986).

Tam, A. C., *Rev. Mod. Phys.* **58**(2), 381 (1986).

6

Applications: Rare Earth and Transition Metal Ions, and Color Centers

6.1 INTRODUCTION

In the previous chapter we have introduced the physical basis of the interpretation of optical spectra of centers in crystals. The main effect of these centers is to introduce new energy levels within the energy gap of the crystal, so that the transitions among these levels produce new optical bands that are not present in the perfect crystal. Due to these absorption and emission bands, centers in crystals are relevant for a variety of applications, such as solid state lasers, amplifiers and phosphors for fluorescent lighting and cathode ray tubes. In this chapter, we will describe the main characteristics of the relevant centers for these applications.

The majority of applications are based on the incorporation of foreign ions in crystals. Regarding our reference model AB_6 (Figure 5.1), this means that A is a foreign ion and B represents the host ligand ions. The foreign ions are deliberately incorporated in the host crystal during the growth processes or, eventually, by other procedures, such as diffusion or ion implantation.

In principle, all of the elements of the periodic table can be used to incorporate foreign ions in crystals. Actually, only a number of elements have been used for optically active centres in crystals; in other words, only a number of elements can be incorporated in ionic form and give rise to energy levels within the gap separated by optical energies. The most relevant centers for technological applications (although not the unique ones) are based on ions formed from the transition metal and rare earth series of the periodic table, so we will focus our attention on these centers.

An Introduction to the Optical Spectroscopy of Inorganic Solids   J. García Solé, L. E. Bausá, and D. Jaque
© 2005 John Wiley & Sons, Ltd   ISBNs: 0-470-86885-6 (HB); 0-470-86886-4 (PB)

Moreover, the analysis of the optical spectra of transition metal and rare earth ions is very illustrative, as they present quite different features due to their particular electronic configurations: transition metal ions have optically active unfilled outer 3d shells, while rare earth ions have unfilled optically active 4f electrons screened by outer electronic filled shells. Because of these unfilled shells, both kind of ion are usually called *paramagnetic ions*.

Optically active centers may also occur as a result of structural defects. These defects are usually called *colour centers*, and they produce optical bands in the colorless perfect crystal. We will also discuss the main features of color centers in this chapter (Section 6.5). From the practical viewpoint, color centers are used to develop solid state lasers. Moreover, the interpretation of their optical bands is also interesting from a fundamental point of view, as these centers can be formed unintentionally during crystal growth and so may give rise to unexpected optical bands.

Finally, in the last section of this chapter (Section 6.6), we will treat two aspects that are of great relevance in the optical spectroscopy of solids. First, we will introduce a semi-empirical method (due to Judd,1962; Ofelt,1962) that analyzes the absorption spectra of trivalent rare earth ions in crystals to search for new efficient phosphors and solid state lasers. Secondly, we will treat a relatively new topic related to optical centers in solids: the optically induced cooling of trivalent ytterbium doped solids.

6.2 RARE EARTH IONS

The rare earth (RE) ions most commonly used for applications as phosphors, lasers, and amplifiers are the so-called *lanthanide ions*. Lanthanide ions are formed by ionization of a number of atoms located in periodic table after lanthanum: from the cerium atom (atomic number 58), which has an outer electronic configuration $5s^2 5p^6 5d^1 4f^1 6s^2$, to the ytterbium atom (atomic number 70), with an outer electronic configuration $5s^2 5p^6 4f^{14} 6s^2$. These atoms are usually incorporated in crystals as divalent or trivalent cations. In trivalent ions 5d, 6s, and some 4f electrons are removed and so $(RE)^{3+}$ ions deal with transitions between electronic energy sublevels of the $4f^n$ electronic configuration. Divalent lanthanide ions contain one more f electron (for instance, the Eu^{2+} ion has the same electronic configuration as the Gd^{3+} ion, the next element in the periodic table) but, at variance with trivalent ions, they tand use to show f $\rightarrow$ d interconfigurational optical transitions. This aspect leads to quite different spectroscopic properties between divalent and trivalent ions, and so we will discuss them separately.

6.2.1 Trivalent rare Earth Ions: the Dieke Diagram

Trivalent lanthanide ions have an outer electronic configuration $5s^2 5p^6 4f^n$, where n varies from 1 (Ce^{3+}) to 13 (Yb^{3+}) and indicates the number of electrons in the unfilled 4f shell. The $4f^n$ electrons are, in fact, the valence electrons that are responsible for the optical transitions.

Table 6.1   The number of 4f electrons
(n) in trivalent lanthanide ions

Ion	n
Ce^{3+}	1
Pr^{3+}	2
Nd^{3+}	3
Pm^{3+}	4
Sm^{3+}	5
Eu^{3+}	6
Gd^{3+}	7
Tb^{3+}	8
Dy^{3+}	9
Ho^{3+}	10
Er^{3+}	11
Tm^{3+}	12
Yb^{3+}	13

Table 6.1 gives the number of 4f valence electrons for each trivalent rare earth ion of the lanthanide series. These valence electrons are shielded by the 5s and 5p outer electrons of the $5s^2 5p^6$ less energetic configurations. Because of this shielding effect, the valence electrons of trivalent rare earth ions are weakly affected by the ligand ions in crystals; a situation that corresponds to the case of a *weak crystalline field*, which was discussed in the previous chapter (Section 5.2). Consequently, the spin–orbit interaction term of the free ion Hamiltonian is dominant over the crystalline field Hamiltonian term. This causes the $^{2S+1}L_J$ states of the $(RE)^{3+}$ ions to be slightly perturbed when these ions are incorporated in crystals. The effect of the crystal field is to produce a slight shift in the energy of these states and to cause additional level splitting. However, the amount of this shift and the splitting energy are much smaller than the spin–orbit splitting, and thus, the optical spectra of $(RE)^{3+}$ ions are fairly similar to those expected for free ions. Moreover, this implies that the main features of a $(RE)^{3+}$ ion spectrum are similar from one crystal to another. In fact, the interpretation of the absorption and luminescence spectra of lanthanide $(RE)^{3+}$ ions in crystals is based on systematic spectral measurements made in a particular host, lanthanum chloride ($LaCl_3$). These spectra were obtained by Dieke and co-workers (1968) and provide a famous energy-level diagram, the so-called *Dieke diagram*, shown in Figure 6.1. This diagram shows the energy of the $^{2S+1}L_J$ states for the $(RE)^{3+}$ ions in $LaCl_3$. The width of each state indicates the magnitude of the crystal field splitting, while the center of gravity of each multiplet gives the approximate location of its corresponding free ion $^{2S+1}L_J$ energy level.

Of course, the energy splitting and center of gravity of the $^{2S+1}L_J$ energy levels for a $(RE)^{3+}$ ion can change slightly when it is incorporated into crystals other than $LaCl_3$, but the gross features of its energy-level diagram remain unchanged. According to its degeneracy, the maximum number of split components for each $^{2S+1}L_J$ multiplet is

Energy (x10³ cm⁻¹)

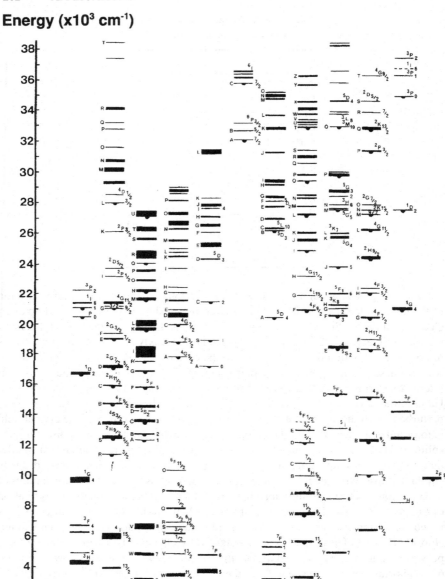

Figure 6.1 An energy-level diagram for trivalent lanthanide rare earth ions in lanthanum chloride (after Dieke, 1968).

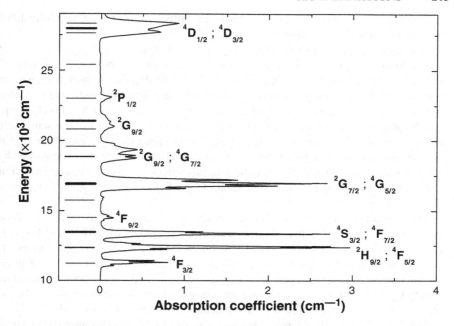

Figure 6.2 The absorption spectrum of Nd^{3+} ions in $LiNbO_3$, taken at room temperature (right-hand side) (registered by the authors). The Dieke diagram levels corresponding to the Nd^{3+} ion are shown on the left-hand side.

$(2J + 1)$ for integer J, or $(J + \frac{1}{2})$ for half-integer J.[1] However, the actual number of components is determined by the local symmetry around the $(RE)^{3+}$ ion in the crystal, as shown in the next chapter.

As an example of how to employ the Dieke diagram to interpret the absorption spectra of trivalent rare earth ions in crystals, we can use the spectrum shown in Figure 6.2. In this figure, the room temperature absorption spectrum of Nd^{3+} ion in the lithium niobate crystal is shown together with the Dieke diagram corresponding to the Nd^{3+} ion. This ion, which is the active ion most frequently used in solid state lasers, shows several groups of sharp optical absorption bands (lines). These lines correspond to transitions from sublevels of the ground state, $^4I_{9/2}$, to sublevels of the different $^{2S+1}L_J$ excited states. These sublevels appear as a result of the splitting due to the crystal field acting on the Nd^{3+} ions in lithium niobate (Tocho *et al.*, 1991). With the help of the Dieke diagram (the left-hand part in Figure 6.2), we can properly assign the different groups of lines to specific $^{2S+1}L_J$ states. Although in some cases overlap between different states is observed, this is an important step in the subsequent identification and labeling of the energy sublevels responsible for the different lines in the spectrum. If the absorption spectra are registered at low temperature, so that only the lowest energy sublevel (0 cm^{-1}) of the ground state is populated, the different

---

[1] This is due to the so-called *Kramers theorem*, which establishes that all electronic energy levels containing an odd number of electrons are at least doubly degenerate.

sublevels of the excited states can be easily identified, as no overlap with transitions from other ground state sublevels can take place.

The Dieke diagram also provides very useful information with which to predict and/or to make a proper assignment of the emission spectra corresponding to trivalent rare earth ions in crystals. Reexamining the Dieke diagram (Figure 6.1), we can observe that some energy states are marked with a semicircle below them. Those marked states correspond to *light-emitting levels* (levels from which a direct depopulation produces luminescence). On the other hand, energy levels with no semicircle below them are those from which direct emission of light has not been observed (of course, they can give rise to luminescence, but from lower energy states populated by nonradiative relaxation). The explanation for this will be given in the next section. For the moment, we can use the Dieke diagram to predict and explain emission spectra from emitting levels.

As an example, Figure 6.3 shows an emission spectrum of the Eu^{3+} ion in the same previous host matrix, $LiNbO_3$. This rare earth ion (together with Tb^{3+} and Eu^{2+}) is commonly used in the fabrication of phosphors for red, green, and blue lamps. Its luminescence spectrum produces red radiation and consists of four groups of lines. These lines correspond to different transitions from the singly degenerate excited state $^5D_0$ to the different sublevels within four terminal states ^7F_J (from $J = 1$ to 4), the $^5D_0 \rightarrow {}^7F_0$ emission is forbidden at the electric dipole order and is not observed. For instance, the $^5D_0 \rightarrow {}^7F_1$ emission transition shows two peaks, because the terminal level $^7F_1$ splits into two levels (one singly degenerate and the other doubly degenerate). Similar analyses can be carried out for the remaining emission peaks related to the other $^5D_0 \rightarrow {}^7F_J$ transitions, thus yielding the splitting components of the other ^7F_J terminal levels.

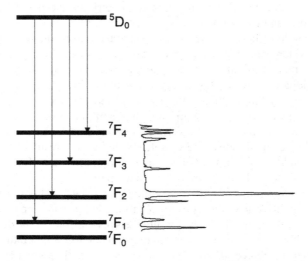

Figure 6.3   The low-temperature emission spectrum of Eu^{3+} ions in $LiNbO_3$. Part of the Dieke diagram for the Eu^{3+} ion is included for explanation (reproduced with permission from Muñoz *et al.*, 1995).

In general, the Dieke diagram can be used as a guide to roughly predict the average wavelength for each $^{2S+1}L_J \rightarrow {}^{2S'+1}L'_{J'}$ transition of a given trivalent rare earth ion in any crystal host. Conversely, it can be used to assign a particular $^{2S+1}L_J \rightarrow {}^{2S'+1}L'_{J'}$ transition to any absorption/emission group of lines, as shown in the next example.

EXAMPLE 6.1   *If a phosphor emitting around 285 nm is to be fabricated based on a certain Tm^{3+} ion doped compound, determine the transition that produces emission around this wavelength and the maximum number of expected peaks.*

The wavenumber of the 285 nm emitted photons is $35\,088$ cm^{-1}. Examining the Dieke diagram column corresponding to Tm^{3+} ions in Figure 6.1, we observe that that the energy matches the energy difference between the $^3P_0$ excited state and the $^3H_6$ ground state. Thus, the transition that should operate at about 285 nm is the $^3P_0 \rightarrow {}^3H_6$ transition.

The excited state $^3P_0$ is singly degenerate, while the ground state $^3H_6$ has a degeneracy of $2J + 1 = 13$. Therefore, the maximum number of emission peaks that we could observe around 285 nm would be 13. Of course, this would occur only in crystals leading to a very low symmetry crystal field around the Tm^{3+} ions. The actual number of peaks depends on the crystal and, specifically, on the crystal field symmetry around the Tm^{3+} ions in this crystal.

Finally, we should remember that f $\rightarrow$ f transitions are parity-forbidden. However, most of them become partially allowed at the electric dipole order as a result of mixing with other orbitals that have different parity because of a noninversion symmetry crystal field (see Section 5.3). Thus, a proper choice of the crystal host (or the site symmetry) can cause a variety of $(RE)^{3+}$ transitions to become *forced electric dipole* transitions.

6.2.2 Divalent Rare Earth Ions

Divalent rare earth ions also have an outer electronic configuration of $4f^n$ (n including one more electron than for the equivalent trivalent rare earth). However, unlike that of $(RE)^{3+}$ ions, the $4f^{(n-1)}5d$ excited configuration of divalent rare earth ions is not far from the $4f^n$ fundamental configuration. As a result, $4f^n \rightarrow 4f^{(n-1)}5d$ transitions can possibly occur in the optical range for divalent rare earth ions. They lead to intense (parity-allowed transitions) and broad absorption and emission bands.

As a relevant example, Figure 6.4 shows the room temperature absorption spectrum of Eu^{2+} in sodium chloride (NaCl). In this crystal, europium is incorporated in the divalent state, replacing Na^+ lattice ions. The spectrum of Eu^{2+} ion in NaCl consists of two broad bands, centered at about 240 nm and 340 nm, which correspond to transitions from the ground state ($^8S_{7/2}$) of the $4f^7$ electronic configuration to states of the $4f^6\,5d$ excited electronic configuration. In fact, the energy separation between

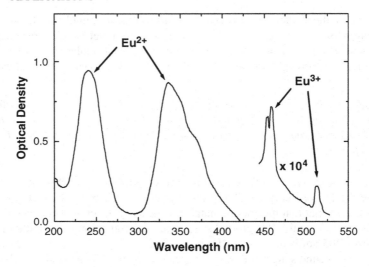

Figure 6.4 The absorption spectrum of Eu^{2+} ions in NaCl. A weak absorption signal related to the Eu^{3+} ions is observed after X-ray irradiation (note the change in the absorption scale) (reproduced with permission from Aguilar *et al.*, 1982).

these two absorption bands is a measure of the octahedral crystal field acting on the Eu^{2+} ions. This crystal field produces a splitting (of amount $10Dq$) of the 5d excited orbital into its t_{2g} and e_g components, as described in Section 5.2. The structure observed in the low-energy band is due to the interaction of the 5d electron with the $4f^6$ electrons.

After X-ray irradiation of thermally annealed NaCl crystals, a small percentage of divalent europium ions are converted into trivalent europium ions (Aguilar *et al.*, 1982). This is shown by the appearance of weak and narrow absorption lines at around 460 nm and 520 nm, related to the $^7F_0 \rightarrow {}^5D_2$ and $^7F_0 \rightarrow {}^5D_1$ transitions of Eu^{3+} ions, respectively. For our purposes, this example allows us to compare the different band features between $(RE)^{3+}$ and $(RE)^{2+}$ ions; Eu^{2+} ions show broad and intense optical bands (electric dipole allowed transitions), while Eu^{3+} ions present narrow and weak optical lines (forced electric dipole transitions).

The absorption and emission bands of Eu^{2+} ions show strong changes from one host crystal to another. For instance, Eu^{2+} ions emit in the violet region when incorporated into the $Sr_2P_2O_7$ compound, while they emit in the green region in $SrAl_2O_4$ phosphor. This behavior is opposite to the practical insensitivity of the absorption and emission spectra of the trivalent europium ion to different host crystals.

6.3 NONRADIATIVE TRANSITIONS IN RARE EARTH IONS: THE 'ENERGY-GAP' LAW

In the Dieke diagram, which is used to interpret the spectra of trivalent rare earth ions (Figure 6.1), we emphasized the emitting levels by marking a semicircle below them.

An inspection of this diagram also shows a variety of *nonlight-emitting levels*. From these levels, the nonradiative rate is clearly dominant over the radiative rate and there is no direct emission. The probability of direct radiative emission from a particular excited energy level of a $(RE)^{3+}$ ion is strongly related to the energy separation between this level and the level just below it. We call this energy separation the *energy gap*. An inspection of the Dieke diagram reveals to us that, in general, from energy levels with a low energy gap the de-excitation is mostly nonradiative, while energy levels with a large energy gap are light-emitting levels. In fact, this aspect was expected from expression (5.17), as the radiative rate between two energy levels, A, varies with ω_0^3 (i.e., with the cubic power of the energy gap, $\hbar\omega_0$).

The nonradiative rate, A_{nr}, from a $(RE)^{3+}$ ion level is also strongly related to the corresponding energy gap. Systematic studies performed over different $(RE)^{3+}$ ions in different host crystals have experimentally shown that the *rate of phonon emission*, or *multiphonon emission rate*, from a given energy level decreases exponentially with the corresponding energy gap. This behavior can be expressed as follows:

$$A_{nr} = A_{nr}(0)e^{-\alpha\Delta E} \tag{6.1}$$

where $A_{nr}(0)$ and α are constants that depend on the host material, but not on the trivalent rare earth ion, and ΔE is the energy gap. This experimental law is known as the *energy-gap law* and it is experimentally manifested when using a low ion doped crystal. The energy-gap law is clearly apparent in Figure 6.5, which shows the experimentally determined values of A_{nr} from different energy levels, of different $(RE)^{3+}$

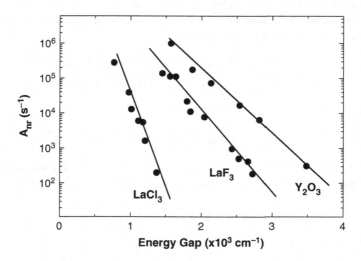

Figure 6.5   The measured values of the nonradiative rate, A_{nr} as a function of the energy gap for different trivalent rare earth ions in three host crystals. The straight lines correspond to the best fits to Equation (6.1) (note the log scale on the A_{nr} axis) (reproduced with permission from Riseberg and Weber, 1975).

ions, as a function of their corresponding energy-gap values in $LaCl_3$, LaF_3, and Y_2O_3. For each one of these materials, the nonradiative rate decreases exponentially (note the log scale) with the energy gap, according to Equation (6.1). It is important to recall that the energy-gap law gives the nonradiative rate from each $(RE)^{3+}$ ion level of a given material through knowledge of its energy gap only, independent of either the type of ion or the nature of the electronic emitting energy level. This aspect, which is a consequence of the characteristic weak ion–lattice interaction of $(RE)^{3+}$ ions, allows us to predict the nonradiative rates from energy levels of different $(RE)^{3+}$ ions by using only the Dieke diagram, as in the next example.

EXAMPLE 6.2   *Determine the nonradiative rates from the following energy levels of different ions in lanthanum chloride:* $^4F_{3/2}$ *(Er*$^{3+}$*),*$^3P_0$ *(Pr*$^{3+}$*) and* $^2F_{5/2}$ *(Yb*$^{3+}$*).*

The experimentally obtained energy-gap law for $LaCl_3$ has been represented in Figure 6.5. From this figure, one can obtain that the best fit of the experimental data (the black points) to expression (6.1) corresponds to $\alpha = 0.015$ cm and $A_{nr}(0) = 4.22 \times 10^{10}$ s^{-1}. Thus, we can write the energy-gap law for the $LaCl_3$ crystal as follows:

$$A_{nr}(s^{-1}) = 4.22 \times 10^{10} \, e^{-0.015 \Delta E (cm^{-1})}$$

From this expression, we can estimate the multiphonon emission nonradiative rate, A_{nr}, from any particular energy level by simply knowing the energy distance to the next lower energy level (the energy gap), ΔE.

From the Dieke diagram of Figure 6.1, the energy-gap values can be determined for each energy level, giving

$$\Delta E \cong 343 \text{ cm}^{-1} \quad \text{for the } {}^4F_{3/2} \text{ state of } Er^{3+}$$
$$\Delta E \cong 3657 \text{ cm}^{-1} \quad \text{for the } {}^3P_0 \text{ state of } Pr^{3+}$$

and

$$\Delta E \cong 9943 \text{ cm}^{-1} \quad \text{for the } {}^2F_{5/2} \text{ state of } Yb^{3+}$$

Using these ΔE values and the previous expression for the energy-gap law, the following nonradiative rates are obtained:

$$A_{nr}(^4F_{3/2}, Er^{3+}) \cong 2.3 \times 10^8 \text{ s}^{-1}$$
$$A_{nr}(^3P_0, Pr^{3+}) \cong 3.5 \times 10^{-14} \text{ s}^{-1}$$
$$A_{nr}(^2F_{5/2}, Yb^{3+}) \cong 9.7 \times 10^{-56} \text{ s}^{-1}$$

The exponential decrease in the multiphonon emission rate with an increasing energy gap, given by Equation (6.1), is due to an increase in the number of emitted phonons as the energy gap is increased. In fact, the larger the energy gap, the larger is the number of phonons needed to bridge this gap, and so the higher is the order of the perturbation process. At the same time, we know that the higher the order of the perturbation process, the smaller is the de-excitation probability due to multiphonon emission. By the same reasoning, the phonons that are expected to participate in the nonradiative de-excitation process are the highest-energy phonons with an appreciable density of states. These active phonons, which are responsible for the nonradiative de-excitation, are usually called *effective phonons*. Thus, the number of effective phonons, p, involved in a multiphonon emission process from a given energy level is found by dividing the corresponding energy gap by the effective phonon energy: $p = \Delta E / \hbar \Omega$. This allows us to rewrite the energy-gap law (Equation 6.1) in terms of p and the effective phonon energy:

$$A_{nr} = A_{nr}(0) \times e^{-(\alpha \hbar \Omega)p} \tag{6.2}$$

Figure 6.6 shows A_{nr} versus the number of effective phonons, p, for the same three materials of Figure 6.5. The energy of the effective phonons for each host crystal is indicated in the figure caption. An exponential decrease in the nonradiative rate with

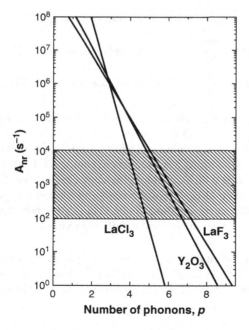

Figure 6.6 The multiphonon nonradiative rate of $(RE)^{3+}$ ions as a function of the number of emitted effective phonons for $LaCl_3$ (260 cm^{-1}), LaF_3 (350 cm^{-1}), and Y_2O_3 (430–550 cm^{-1}). The numbers in brackets indicate the energies of the effective phonons. The shaded area indicates the range of typical radiative rates.

the number of effective phonons is clearly observed for all three crystals, in accordance with Equation (6.2). The shaded area in Figure 6.6 indicates the range (from about 10^2 s^{-1} to about 10^4 s^{-1}) of typical radiative decay rates of $(RE)^{3+}$ ions in crystals. For instance, we can see that, for LaCl$_3$, nonradiative processes are clearly dominant for processes involving less than four effective phonons. On the other hand, de-excitation from energy levels that require more than about five effective phonons occurs mainly by luminescence. Trends in new efficient luminescent materials based on $(RE)^{3+}$ ions are directed toward the search for low effective phonon energy host materials.

EXAMPLE 6.3 *Determine the number of effective phonons involved in the de-excitation processes from the energy levels of the $(RE)^{3+}$ ions in LaCl$_3$ given in Example 6.2.*

Considering that the effective phonons in LaCl$_3$ are those with an energy of 260 cm^{-1} (see the caption to Figure 6.6), and considering the different energy gaps from the $^4F_{3/2}$ (Er^{3+}), $^3P_0$ (Pr^{3+}), and $^4F_{5/2}$ (Yb^{3+}) energy levels, we determine the number, p, of effective phonons in each nonradiative de-excitation process:

$$p = \frac{343}{260} \approx 1 \text{ phonon from the } {}^4F_{3/2} \text{ (Er}^{3+}\text{) state}$$

$$p = \frac{3657}{260} \approx 14 \text{ phonons from the } {}^3P_0 \text{ (Pr}^{3+}\text{) state}$$

$$p = \frac{9943}{260} \approx 38 \text{ phonons from the } {}^2F_{5/2} \text{ (Yb}^{3+}\text{) state}$$

The very low multiphonon decay rates obtained in Example 6.2 from the $^3P_0$ (Pr^{3+}) and $^2F_{5/2}$ (Yb^{3+}) states are due to the large number of effective phonons that need to be emitted – 14 and 38, respectively – and so the high-order perturbation processes. As a consequence, luminescence from these two states is usually observed with a quantum efficiency close to one. On the other hand, from the $^4F_{3/2}$ state of Er^{3+} ions the energy needed to bridge the short energy gap is almost that corresponding to one effective phonon: hence depopulation of this state to the next lower state is fully nonradiative.

Finally, it is important to recall that the simple nonradiative rate law described by Equations (6.1) and (6.2) is only valid for $(RE)^{3+}$ ions. This is a consequence of the weak ion–lattice interactions for these ions, that leads to a Huang–Rhys parameter of $S \approx 0$.

6.4 TRANSITION METAL IONS

Transition metal (TM) ions are frequently used as optically active dopants in commercial phosphors and in tunable solid state lasers. TM ions are formed from atoms in the

Table 6.2   The most common transition
metal ions and their corresponding
numbers (n) of 3d valence electrons

Ions	n
Ti^{3+}, V^{4+}	1
V^{3+}, Cr^{4+}, Mn^{5+}	2
V^{2+}, Cr^{3+}, Mn^{4+}	3
Cr^{2+}, Mn^{3+}	4
Mn^{2+}, Fe^{3+}	5
Fe^{2+}, Co^{3+}	6
Fe^+, Co^{2+}, Ni^{3+}	7
Co^+, Ni^{2+}	8
Ni^+, Cu^{2+}	9

fourth period of the periodic table; from beyond the calcium atom (element 20 in the periodic table), with electronic configuration $(Ar)4s^2$, up to the zinc atom (element 30), with electronic configuration $(Ar)3d^{10}4s^2$. TM atoms tend to lose the outer 4s electrons, and in some cases lose or gain 3d electrons, to form different kinds of stable cations. Thus, TM ions have an electronic configuration $1s^2 2s^2 2p^6 3s^2 3p^6 3d^n$, where n ($1 < n < 10$) denotes the number of 3d electrons. These electrons are responsible for the optical transitions (i.e., they are valence electrons). Table 6.2 lists the most common transition metal ions and the corresponding numbers of 3d electrons.

The 3d orbitals in TM ions have a relatively large radius and are unshielded by outer shells, so that strong ion–lattice coupling tend to occur in TM ions. As a result, the spectra of TM ions present both broad ($S > 0$) and sharp ($S \approx 0$) bands, opposite to the spectra of $(RE)^{3+}$ ions, discussed in section 6.2.1, which only showed sharp bands ($S \approx 0$).

6.4.1 3d^1 Ions

The simplest transition metal ions, such as Ti^{3+} and V^{4+}, have a $3d^1$ outer electronic configuration. The spectra of these ions in an octahedral environment (such as that of Figure 5.1) were discussed in Section 5.2. As a relevant example, Figure 6.7 shows the absorption and emission spectra of Ti^{3+} ions in Al_2O_3(sapphire). Both spectra correspond to transitions between the t_{2g} (ground) and e_g (excited) energy levels, which arise from the splitting of the $3d^1$ level caused by the octahedral crystal field. The bands are broad and there is a large Stokes shift between absorption and emission, in accordance with a strong electron–lattice coupling that leads to a large Huang–Rhys S value. In addition, the $t_{2g} \leftrightarrow e_g$ transitions are electric dipole allowed, because of the crystal field around the Ti^{3+} ions in Al_2O_3, which is slightly distorted from a perfect octahedral environment and so produces mixing of configurations with opposite parities. Thus, the radiative lifetime is relatively short (3.9 μs), typical of

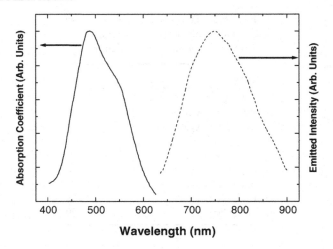

Figure 6.7 Room temperature absorption (full line) and emission (dashed line) spectra of Ti^{3+} ions in Al_2O_3 (sapphire) (reproduced with permission from Moulton, 1983).

an electric dipole forced transition. Moreover, the luminescence efficiency is still high at room temperature. Because of their high efficiency and broadband features, Ti–sapphire crystals are excellent for the development of broadly tunable solid state lasers in the red – near infrared spectral region, as mentioned in Section 2.5.

6.4.2 $3d^n$ Ions: Sugano–Tanabe Diagrams

The case of ions with more than one 3d electron is more complicated, as the interactions among these valence electrons, H_{ee} in Equation (5.4), must be taken into account. In general, the problem is solved by perturbative methods, considering that both H_{ee} and the crystal field Hamiltonian, H_{CF}, are dominant over the spin–orbit interaction, H_{SO}. This perturbation approach is a consequence of the Z^4 dependence (Z being the atomic number) of the spin–orbit interaction and the not very large Z values of the TM ions (from 21 to 30). This behavior is at variance with the case of RE ions, where Z varies from 58 (Ce) to 70 (Yb), and so the spin–orbit interaction must be considered before the crystal field interaction. The free ion states of transition metal ions are governed by the electron–electron interaction and so are labeled by ^{2S+1}L states (usually called L–S terms), where S is the total spin and L is the total angular momentum. The energy separation between the various ^{2S+1}L states is usually given in terms of the so-called *Racah parameters* (A, B, and C). These parameters describe the strength of the electrostatic interactions between the electrons (Henderson and Imbusch, 1989).

Sugano and Tanabe have calculated the energy of the states deriving from the $3d^n$ ions (from $n = 2$ to $n = 8$) as a function of the octahedral crystal field strength.

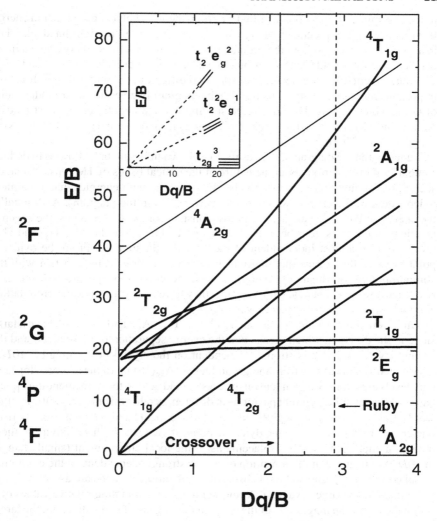

Figure 6.8 The Sugano–Tanabe diagram of a 3d³ electronic configuration ($C/B = 4.5$). The two vertical lines correspond to the Dq/B values in the crossover and for the ruby crystal (see the text). The inset shows the energy-level diagram for strong crystal fields.

These calculations are represented in the so-called *Sugano–Tanabe diagrams*, which are extremely useful in the interpretation of the spectra of TM ions in a variety of host crystals. Sugano–Tanabe diagrams show how the ^{2S+1}L free ion levels split up as the ratio between the crystal-field strength and the interelectronic interaction (a ratio measured in units of Dq/B) increases. Figure 6.8 shows the Sugano–Tanabe diagram for a 3d³ transition metal ion (such us Cr^{3+}, V^{2+}, or Mn^{4+}). The free ion energy levels, ^4F, ^4P, ^2G, and ^2F, are shown on the left-hand side, measured in E/B energy units. The split components (identified by group theory labels) of each free ion energy level in an increasing octahedral crystal field are shown on the right-hand side. For instance, it can be seen how the ground state ^4F splits into three energy levels in an octahedral crystal field; the $^4A_{2g}$ ground level and two excited levels, $^4T_{2g}$ and $^4T_{1g}$. The other excited free ion levels also split into different A T, and E levels. It is also important to mention that a Sugano–Tanabe diagram is given for a particular value of C/B, which mainly depends on the specific ion and slightly on the host matrix. Among the TM free ions, the value of C/B varies from 4.19 for Ti^{2+} to 4.88 for Ni^{2+} (Sugano et al., 1970).

Sugano–Tanabe diagrams, such as the one shown in Figure 6.8, allow us to deduce some useful information about the nature of the optical bands of TM ions. As can be seen in this figure, there are two levels, ^2E_g and $^2T_{1g}$, with energies that are almost independent of the crystal field (close to zero slopes in the diagram). Additionally, the energy of the $^2T_{2g}$ state is also almost constant for $Dq/B > 1$. Thus, the spectral positions of the transitions between the $^4A_{2g}$ ground level and the $^2E_g, ^2T_{1g}$, and $^2T_{2g}$ levels are also almost independent of the crystal field strength. From the dynamic point of view, this means that the transition energy is practically constant with the configurational coordinate Q. Therefore, these close to zero slope energy levels give rise to narrow optical bands, with $S \approx 0$, as they have nearly the same electron–lattice coupling behavior as the ground level ($^4A_{2g}$).

On the other hand, other levels, such as $^4T_{1g}$, $^4T_{2g}$, $^2A_{1g}$, and $^4A_{2g}(^2F)$, have a large slope in the diagram, which means that the energy separation of these levels and the ground level, $^4A_{2g}(^4F)$, is strongly dependent on the crystal field strength,[2] $10Dq$. Consequently, transitions from the ground level $^4A_{2g}(^4F)$ up to these large-slope energy levels are strongly dependent on the crystal field, and so the corresponding optical bands appear at quite different positions in different octahedral environments. In other words, this means that, for a given $3d^3$ ion, these optical bands change strongly from one crystal to another. From the dynamic point of view, the high sensitivity of these transition energies to small displacements of the local environment (small changes in $10Dq$) indicates that the transition energy is strongly dependent on the configurational coordinate Q, and so broad absorption and emission bands are associated with these transitions (large S values). A rule of thumb for estimating relative values of S is that the larger the magnitude of the slope in the Sugano–Tanabe diagram, the larger is the value of S.

The inset in Figure 6.8 shows the splitting of the free ion energy levels in the case of a strong crystal-field approximation. In this case, the electrostatic interactions among the 3d electrons are neglected in comparison to the crystal field. As a result, the orbitals are products of three single electron orbitals, each single orbital being

---

[2] The origin of a different slope for two states lies in the different participation of the e_g and t_{2g} orbitals in these two states (see the paragraph below, related to the strong crystal field case). For instance, the ^2E_g and $^4A_{2g}$ (ground) states derive from t_{2g}^3 orbitals, so they have similar phonon coupling (or $S \approx 0$) and therefore the ^2E_g state has a slope close to zero in the Sugano–Tanabe diagram. On the other hand, the $^4T_{2g}$ state derives from $t_{2g}^2 e_g$ orbitals and so there is a strong difference in the phonon coupling in respect to the ground state, $^4A_{2g}$.

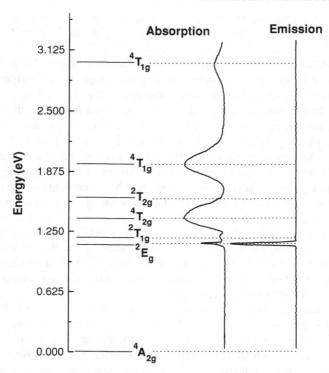

Figure 6.9 The absorption and emission spectra of ruby (Al_2O_3:Cr^{3+}). On the left-hand side, the energy levels of Cr^{3+} in Al_2O_3 are shown for a proper assignment of the observed optical bands.

either a t_{2g} or an e_g orbital. Then, four different combinations are possible: t_{2g}^3, ($t_{2g}^2 e_g^1$), ($t_{2g} e_g^2$), and e_g^3. Among these orbitals, t_{2g}^3 has the lowest energy, the next upper level is ($t_{2g}^2 e_g^1$), and so on, as shown in the inset of Figure 6.8.

As an example of how the Sugano–Tanabe diagram is used to interpret spectra, Figure 6.9 shows the absorption and emission spectra of Cr^{3+} in Al_2O_3 (the ruby laser crystal) together with the energy levels related to these spectra. The energies of the different crystal field levels depend on the parameters Dq, C, and B. These parameters are obtained by comparing the experimental energy levels with the calculated energies (those in the Sugano–Tanabe diagram). For ruby, $Dq/B = 2.8$ and $B = 918$ cm^{-1}. Thus, the dashed vertical line displayed in Figure 6.8, at $Dq/B = 2.8$, gives the positions of the different energy levels of Cr^{3+} in Al_2O_3. The lowest energy transitions are the $^4A_{2g} \rightarrow {}^2E_g, {}^2T_{1g}$ spin-forbidden transitions. As the $^2E_g, {}^2T_{1g}$ excited levels have a close to zero slope on the Sugano–Tanabe diagram, the previous transitions give rise to two sharp absorption bands (note that the $^4A_{2g} \rightarrow {}^2T_{1g}$ absorption is very weak). On the other hand, the $^4A_{2g} \rightarrow {}^4T_{2g}, {}^4T_{1g}$ spin-allowed transitions produce strong and broad absorption bands, this last feature being due to the large slope of the terminal levels on the Sugano–Tanabe diagram (see Figure 6.8). The two broad absorption bands in the yellow–green ($^4A_{2g} \rightarrow {}^4T_{2g}$) and in the blue ($^4A_{2g} \rightarrow {}^4T_{1g}$) are responsible for the red color of ruby crystals.

Although the absorption spectrum of ruby shows several broad and sharp bands, the luminescence spectrum is much more simple as, regardless the excitation wavelength, it only consists of a sharp emission related to the $^2E_g \rightarrow {}^4A_{2g}$ lowest energy transition.[3] In fact, the generation of radiative emission from only the first excited level is a common feature of transition metal ions. This is because TM ions have a more complex ion–lattice interaction than $(RE)^{3+}$ ions and the multiphonon decay rate of TM ions does not follow the energy-gap law given by Equations (6.1) and (6.2). In any case, there is still some qualitative validity in the energy-gap rule: a nonradiative process that involves no more than 20–25 phonons is, in general, more efficient than a radiative process across the same energy gap. In other words, this means that nonradiative decay processes are more favorable for TM ions than for $(RE)^{3+}$ ions. This is due to the appearance of several bands with $S \gg 0$ for TM ions, which makes possible the depopulation of their excited states by real phonon energy levels, as described in Section 5.6. Therefore, excitation into any excited state of a TM ion is followed by nonradiative multiphonon emission down to the lowest excited level, from which the emission originates.

Another important detail on the Sugano–Tanabe diagram shown in Figure 6.8 is the vertical line at the value $Dq/B = 2.2$, at which the states $^4T_{2g}$ and ^2E are equal in energy. This value of Dq/B is usually referred to as the *crossover value*. Materials (crystal + ion) for which Dq/B is less than the crossover value are usually called *low crystal field materials*. For these materials, the lowest energy level is the $^4T_{2g}$ and so they present a characteristic broad and intense emission band associated with the spin-allowed $^4T_{2g} \rightarrow {}^4A_{2g}$ transition (which is usually a vibronic transition). On the other hand, the materials on the right-hand side of the crossover line are called *high crystal field materials*. These materials (such as the ruby crystal) present a narrow-line emission related to the spin-forbidden $^2E_g \rightarrow {}^4A_{2g}$ transition, usually called *R-line emission*.

Figure 6.10 shows the emission spectra of the Cr^{3+} ion in different host crystals. For each crystal the energy separation, ΔE, between the excited states $^4T_{2g}$ and ^2E_g is indicated. In fact, this amount of energy is another quantitative measure of the crystal field strength; crystals with $\Delta E > 0$ are high crystal field materials, while those for which $\Delta E < 0$ are low crystal field materials. While Cr^{3+} in Al_2O_3 (ruby) shows a sharp R line emission at 694.3 nm, corresponding to a high crystal field ($\Delta E = 2300$ cm^{-1}), the emission spectra of this ion in $BeAl_2O_4$ (alexandrite), $\Delta E = 700$ cm^{-1}, and in $Be_3Al_2(SiO_3)_6$ (emerald), $\Delta E = 400$ cm^{-1}, show both the narrow and broadband emissions, in spite of the fact that both systems are on the high crystal field side. This is because the latter two systems are near the crossover point. For these systems, and taking into account that the lifetime of the ^2E_g state is substantially larger (in the order of ms) than that of the $^4T_2$ state (in the order of

---

[3] In fact, this emission is split into two sharp lines, due to a symmetry that is lower than octahedral symmetry.

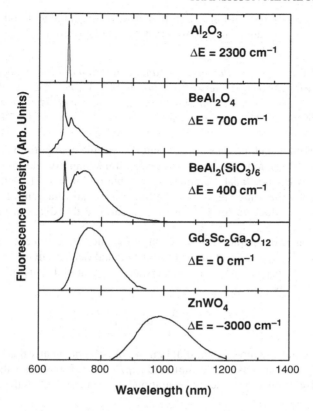

Figure 6.10   The room temperature luminescence of the Cr^{3+} ion in different host crystals. ΔE indicates the energy separation between the $^2T_{2g}$ and ^2E_g excited states (reproduced with permission from Moulton, 1985).

μs), a thermalization process can partially populate the $^4T_2$ level from the ^2E_g level, with the subsequent emission transitions $^2E_g \rightarrow {}^4A_{2g}$ and $^4T_{2g} \rightarrow {}^4A_{2g}$. Therefore, we usually say that alexandrite and emerald systems are *intermediate crystal field* materials. However, at low temperature only the narrow R line emission is observed in intermediate crystal field materials. On the contrary, the emission spectra of Cr^{3+} in $Gd_3Sc_2Ga_3O_{12}$ (GSGG), $\Delta E \approx 0\,cm^{-1}$, and in $ZnWO_4$, $\Delta E = -3000\,cm^{-1}$, only show the broadband $^4T_{2g} \rightarrow {}^4A_{2g}$ emission, typical of low crystal field materials at any temperature.[4] The broadband emissions of the Cr^{3+} ion, displayed in Figure 6.10, have provided the possibility of realizing efficient tunable solid state lasers in the red and infrared based on these host crystals. In addition, it can be seen how, in agreement with the Sugano–Tanabe diagram prediction, the emission spectrum is shifted to lower energies as the crystal field strength (or ΔE) decreases. This allows us to select

---

[4] For these materials, thermalization from the $^4T_{2g}$ to the upper ^2E_g level does not occur, because the lifetime of the $^4T_{2g}$ level is much shorter than the lifetime of the ^2E_g level.

a suitable host crystal to emit in any particular wavelength range. In other words, the Sugano–Tanabe diagram is a guide for crystal field engineering.

EXAMPLE 6.4   *Using the Sugano–Tanabe diagram of Figure 6.8, estimate the energy for the lowest energy transition in $V^{2+}:MgF_2$ ($\Delta E \approx -2500$ cm^{-1}) and infer details about the emission of this laser material. The Racah parameters for the V^{2+} free ions are $B = 755$ cm^{-1} and $C = 3257$ cm^{-1}.*

Divalent vanadium ions have an outer electronic configuration of $3d^3$ (see Table 6.2). These ions occupy a near octahedral symmetry site in magnesium fluoride. Using the Racah parameters, we obtain $C/B = 4.31$. Although this is not exactly the same value as in the Sugano–Tanabe diagram of Figure 6.8 ($C/B = 4.50$, adequate for Cr^{3+} ions), we can use this diagram to make an estimation.

The energy separation between the $^4T_{2g}$ and ^2E_g excited states, $\Delta E \approx -2500$ cm^{-1}, is an indirect measure of the crystal field strength. The negative sign indicates that $V^{2+}:MgF_2$ is a low crystal field material. This value is given, in units of B, by

$$\frac{\Delta E}{B} = -3.31$$

Using the Sugano–Tanabe diagram, we obtain graphically that this energy separation corresponds to a crystal field $Dq/B \approx 1.8$. From this value, we can graphically estimate the energy separation between the first excited state $^4T_{2g}$ and the ground state $^4A_{2g}$:

$$E(^4T_{2g}) - E(^4A_{2g}) \approx 16B = 12\,080 \text{ cm}^{-1}$$

which corresponds to a wavelength of 828 nm.

Indeed, this is not far off the peak wavelength (about 850 nm) experimentally observed for the $^4A_{2g} \rightarrow {}^4T_{2g}$ absorption band (Moulton, 1985). The agreement is good, especially when we take into account that we are using free ion Racah parameters, which can be slightly modified in the crystal, and a Sugano–Tanabe diagram with a nonexact C/B value.

As we are considering a low crystal field material, we expect a broad $^4T_{2g} \rightarrow {}^4A_{2g}$ emission because of the large slope of the $^4T_{2g}$ level in the Sugano–Tanabe diagram. The actual emission of V^{2+} in MgF_2 consists of a broad infrared band, peaking of about 1100 nm. Of course, our simple approach does not take into account the Stokes shift expected for the $^4T_{2g} \rightarrow {}^4A_{2g}$ vibronic transition and so the actual peak of the emission band cannot be predicted.[5]

---

[5] In fact, the 828 nm wavelength obtained for the $^4T_{2g} \rightarrow {}^4A_{2g}$ transition would be an average value between the absorption and emission peak wavelengths within the configurational coordinate model.

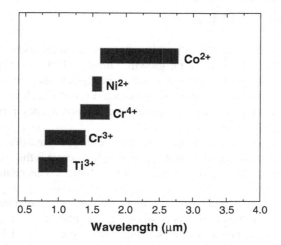

Figure 6.11   The laser wavelength ranges covered by several transition metal ions when they are incorporated in different crystals.

The incorporation of TM ions in different host matrices has led to the development of a great variety of tunable solid state lasers (see Chapter 2). These lasers cover a wide spectral range in the red and in the infrared, as shown in Figure 6.11. They work at room temperature, except for those based on Ni^{2+} and Co^{2+}, which operate at low temperatures (77 K). It is also important to say that Ni^{2+}, Co^{2+}, and Cr^{4+} ions are incorporated in tetrahedral sites in a great variety of crystals, so that the octahedral Sugano–Tanabe diagrams cannot be used to interpret their spectra. In this respect, König and Kremer (1997) have computed similar diagrams for symmetries other than the octahedral one.

The incorporation of Cr^{4+} ions in crystals is presently an active research subject, due to the possibility of realizing new broadly tunable solid state lasers in the infrared, which will operate at room temperature. Moreover, the spectroscopic properties of this ion are particularly useful in the development of saturable absorbers for Q-switching passive devices. At the present time, Cr^{4+}:YAG is the most common material employed as a passive Q-switch in Nd:YAG lasers. This is because the Cr^{4+} ions provide an adequate absorption cross section at the Nd^{3+} laser wavelength (1.06 μm), together with the good chemical, thermal, and mechanical properties of YAG crystals, which are required for stable operation.

Mn^{2+} ions ($3d^5$ electronic configuration) are shown to produce a broad luminescence in more than 500 inorganic compounds, covering a wavelength range from about 490 nm to about 750 nm. Although this ion is not of relevance for laser applications, it is widely used in the phosphor screens of cathode ray tubes and in fluorescent lamps.

Exhaustive information about the applications of optically active ions in phosphors can be found in Shionoya and Yen (1998).

6.5 COLOR CENTERS

So far, we have dealt with optically active centers based on dopant ions, which are generally introduced during crystal growth. Other typical optically active centers are associated with intrinsic lattice defects. These defects may be electrons or holes associated with vacancies or interstitials in ionic crystals, such as the alkali halide matrices. These centers are usually called *color centers*, as they produce coloration in the perfect colorless crystals.

Figure 6.12 shows a schematic representation of the structures of some color centers in an alkali halide crystal, such as NaCl. The simplest center is the so-called *F center*, where *F* stands for *'Farbe'*, the German word for 'color'. This center consists of an electron trapped in the attractive electric field of an anionic vacancy (a Cl^- vacancy in NaCl). Color centers can also be formed by aggregates of two, three, or four single *F* centers in nearest neighbor positions, giving place to F_2, F_3, and F_4 centers, respectively. An F_2 center has been schematically represented in Figure 6.12. The coagulation of color centers can even give rise to the appearance of colloids.

Because of their importance for solid state lasers, the F_A and F_2^+ color centers are, amongst others, of special relevance. The F_A center consists of an *F* center in which one of the neighboring cations (Na^+ in sodium chloride) has been replaced by another cation that is different to that of the host crystal, as shown in Figure 6.12. In F_B centers (not represented in Figure 6.12), two lattice cations close to the *F* center are replaced by other foreign cations. The possibility of the different color centers gaining or losing electrons leads to a new variety of centers. In Figure 6.12,

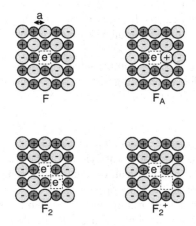

Figure 6.12 The structures of some typical color centers in alkali halide crystals (such as NaCl). The defects are represented on a plane of the alkali halide crystal. The circles represent the lattice ions and 'a' is the anion–cation distance.

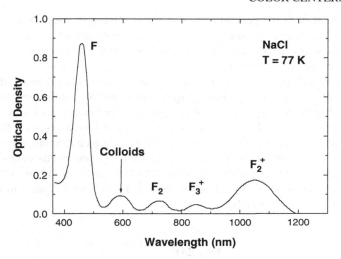

Figure 6.13   The low-temperature (77 K) absorption spectrum of a F-colored NaCl crystal after room temperature aggregation under illumination with light lying in the F absorption band. Bands associated with different color centers can be clearly seen (reproduced with permission from Lifante, 1989).

we have represented the F_2^+ center, in which an electron has been removed from an F_2 center.

Color centers, like those shown in Figure 6.12, give rise to energy levels located within the energy gap of the host crystal and, consequently, to the possible appearance of optical bands as a result of allowed transitions between these energy levels. Figure 6.13 shows the absorption spectrum of a NaCl crystal containing a variety of color centers. Broad absorption bands centered at about 450 nm (F), 580 nm (colloids), 720 nm (F_2), 850 nm (F_3^+), and 1050 nm (F_2^+), associated with different color centers, can be clearly seen. The wide variety of color centers provides great versatility in the production of luminescent systems, which are of particular interest for tunable laser applications.

Usually, the different types of color center are formed after a certain initial concentration of F centers has been produced. These primary centers are typically created by two main experimental methods; (i) *additive coloration* and (ii) *irradiation*.

(i) In *additive coloration*, an alkali halide crystal is heated in an alkali vapor atmosphere to permit its diffusion, and then rapid cooling. As a result, an excess of the alkali metal is introduced and anion vacancies are created to preserve the full charge neutrality. The anion vacancies act like positive holes that can attract electrons to form F centers.

(ii) Alternatively, the mentioned anion vacancies can be produced by *irradiation* with UV, X-rays, γ-rays, or electrons. In these cases, the initial process induced by the irradiation is the generation of electron–hole pairs. The nonradiative energy delivered in the recombination process produces displacements of halogens (usually in an atomic state) toward interstitial positions, so that halogen vacancies with trapped electrons (F centers) are created.

Once a certain concentration of F centers has been reached, the generation of other centers can be carried out by different aggregation and/or illumination processes.

Because of their importance as basic primary centers, we will now discuss the optical bands associated with the F centers in alkali halide crystals. The simplest approximation is to consider the F center – that is, an electron trapped in a vacancy (see Figure 6.12) – as an electron confined inside a rigid cubic box of dimension $2a$, where a is the anion–cation distance (the $Cl^- - Na^+$ distance in NaCl). Solving for the energy levels of such an electron is a common problem in quantum mechanics. The energy levels are given by

$$E_n = \frac{h^2}{8m_0(2a)^2}(n_x^2 + n_y^2 + n_z^2) \tag{6.3}$$

where m_0 is the electronic mass and n_x, n_y, and n_z are quantum numbers that have integer values of $1, 2, 3, \ldots, \infty$. The ground state corresponds to $n_x = n_y = n_z = 1$, while the first excited state has one of these quantum numbers equal to 2 and the others equal to 1. Thus, the lowest energy transition of the F center occurs at a photon energy E_F given by

$$E_F = \frac{3h^2}{8m_0}\frac{1}{(2a)^2} \tag{6.4}$$

EXAMPLE 6.5   *Using the simple model of an electron in a rigid box, as previously described, estimate the wavelength peak for the F absorption band in NaCl (the $Cl^- - Na^+$ distance is 0.28 nm).*

Inserting the values of the constants in Equation (6.4), this can be written in terms of the anion–cation distance as

$$E_F \text{ (eV)} = \frac{1.13}{(2a \text{ (nm)})^2}$$

Substituting $a = 0.28$ nm in this expression, we obtain $E_F = 3.6$ eV $\cong$ 344 nm. This value is, in fact, an overestimation of the actual absorption peak of energy of the F centers in sodium chloride (451 nm).

As shown in Example 6.5, the previous simple model for the F center overestimates its transition energy somewhat. In fact, the experimentally measured values

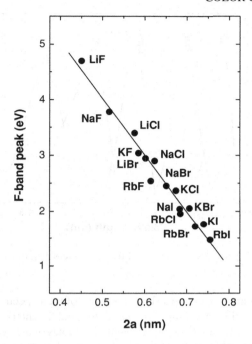

Figure 6.14 Variation of the *F* absorption band peak (note the log scale) with the lattice parameter (2*a*) in alkali halides (reproduced with permission from Dawson and Pooley, 1969).

do not match the $(2a)^{-2}$ variation given in Equation (6.4). Rather, the experimental absorption peaks (see Figure 6.14) follow a best fit to the expression

$$E_F \text{ (eV)} = \frac{0.97}{(2a(\text{nm}))^{1.77}}$$ (6.5)

This disagreement with expression (6.4) is because a real *F* center does not exactly correspond to a cubic box; this is only a rough approximation. Moreover, the *F* center structure is not rigid (a rigid cubic box) but dynamic. In this dynamic structure, the neighboring ions (Na^+ and Cl^- in NaCl) are participating in the lattice vibration modes (phonons). This would produce shifting of the neighboring ions from their equilibrium positions, thus altering the size of the box where the electron is trapped. Hence, we expect a strong electron–phonon coupling for color centers. This is, in fact, experimentally manifested in broad absorption and emission bands and large Stokes shifts (see Exercise 6.11). The Huang–Rhys parameters of the *F* center bands are usually very large, varying from $S = 28$ (NaF) to $S = 61$ (LiCl) in alkali halide crystals. The typical strong electron–phonon interaction of color centers is a great advantage for broadly tunable solid state lasers based on color centers. However, in

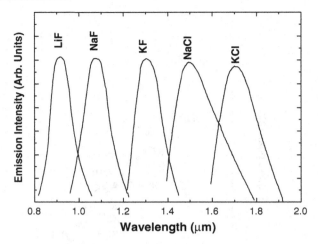

Figure 6.15   The emission spectra of the F_2^+ centers in various alkali halides.

general, F centers are inoperable as lasers because of their spectroscopic parameters, which lead to low gain factors. Nevertheless, other color centers, such as the F_A and F_2^+ centers shown in Figure 6.12, are excellent examples of active centers for tunable solid state lasers. As an example, Figure 6.15 shows the emission spectra of the F_2^+ centers in various alkali halide crystals. It can be seen how an infrared range from about 0.8 μm to almost 2 μm can be covered by an adequate choice of the host crystal for the F_2^+ centers (Lifante, 1989).

In general, the limitations of color centers as lasers arise from their poor thermal stability and the degradation of their optical bands under illumination. In fact, F_2^+ center based solid state lasers must be stabilized by incorporating additional impurities into the crystals. Additionally, color center lasers should operate at low temperature, a problem that strongly limits their manufacture.

6.6 ADVANCED TOPICS: THE JUDD AND OFELT FORMALISM, AND OPTICAL COOLING OF SOLIDS

In this section, we will describe two advanced topics that are directly related to the spectroscopy of trivalent rare earth ions.

In the first topic, we will briefly describe a semi-empirical method that is commonly used to estimate the radiative transition probabilities from energy levels of $(RE)^{3+}$ ions in crystals. This is certainly very useful in order to determine the efficiency of a $(RE)^{3+}$ based system as a luminescent or laser material. In the previous chapter (Section 5.7), we have described a method for determining the quantum efficiency of a luminescent system. However, the application of that method is limited to certain cases and additional approaches become necessary for the description of most $(RE)^{3+}$ ion doped crystals.

In the second topic, we will describe a very smart new application based on Yb^{3+} ion doped glasses: laser cooling of condensed matter.

6.6.1 The Judd and Ofelt Formalism

We have seen that the absorption spectra of $(RE)^{3+}$ ions (see Figures 6.2 and 6.3) consist of several sets of lines corresponding to transitions between the Stark sublevels of $^{2S+1}L_J$ states within the $4f^n$ electronic configuration. A typical absorption spectrum of a $(RE)^{3+}$ ion in crystals is like the one sketched in Figure 6.16. The different sets of transitions correspond to different $J \rightarrow J'$ transitions (J accounting for the ground state), which, in principle, are only permitted at magnetic dipole order: the selection rule is $\Delta J = 0, \pm 1$, with $0 \leftrightarrow 0$ forbidden.

However, although f $\rightarrow$ f transitions are, in principle, forbidden by the Laporte parity rule, most of the transitions in $(RE)^{3+}$ ions occur at the electric dipole (ED) order. As we have already mentioned, this is an ED allowance due to the admixture of the $4f^n$ states with opposite parity excited states $4f^{n-1}5d$, as a result of the lack of inversion symmetry (ED forced transitions). The oscillator strength, f, for a $J \rightarrow J'$ absorption band can be estimated using expression (5.19). We now rewrite this expression as follows:

$$f = \frac{2m\bar{\omega}_0}{3\hbar e^2 (2J+1)} \times |\boldsymbol{\mu}|^2 \qquad (6.6)$$

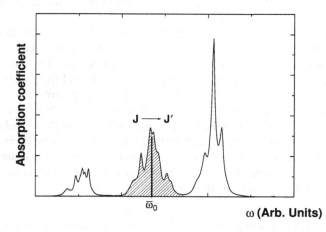

Figure 6.16 A typical absorption spectrum of a trivalent rare earth ion (not corresponding to any specific ion) in a crystal. A generic $J \rightarrow J'$ absorption transition, with an average frequency of ϖ_0, has been marked and shaded (see the text).

where we have included the factor $(2J + 1)$ to take into account the degeneracy of the initial state and an average frequency of $\overline{\omega}_0$ corresponding to the $J \rightarrow J'$ transition (see Figure 6.16).

Judd (1962) and Ofelt (1962) demonstrated that, for electric dipole f $\rightarrow$ f transitions and under certain approximations,[6] the square of the matrix element in Equation (6.6) can be written as follows:

$$|\mu|^2 = e^2 \sum_{t=2,4,6} \Omega_t \times \left| \left\langle \alpha J \left\| U^{(t)} \right\| \alpha' J' \right\rangle \right|^2 \tag{6.7}$$

where the $\Omega_t (t = 2, 4,$ and 6) are the so-called *Judd–Ofelt intensity parameters* and the $|\langle \alpha J \| U^{(t)} \| \alpha' J' \rangle|$ are the reduced matrix elements of tensor operators of rank t (α and α' symbolizing all of the quantum numbers needed to the define the J and J' states). The Judd–Ofelt Ω_t parameters characterize the strength and the nature of the odd-parity crystal field acting on the $(RE)^{3+}$ ion. The reduced matrix elements in Equation (6.7) can be calculated from published data and are generally independent of the host crystal. They have, in fact, been tabulated by (Carnall *et al.*, 1968). This implies that the oscillator strength for any absorption or emission transition can be determined (by expression 6.7) if the Ω_t parameters are known.

The Judd–Ofelt intensity parameters can be obtained by analyzing the room temperature absorption spectrum of the $(RE)^{3+}$ ion doped crystal, provided that the concentration of these ions and the refractive index of the crystal are known. Using Smakula's formula (5.22), the oscillator strength, f_{exp}, for each $J \rightarrow J'$ absorption transition can be experimentally determined, just by measuring its corresponding absorption area (for instance, the shaded area in Figure 6.16). Then, by a least squares fit of the f_{exp} values to the corresponding calculated values, f_{cal} (using Equations (6.6) and (6.7)), for different $J \rightarrow J'$ absorption transitions, the Ω_t parameters can be determined.

The main advantage of the Judd–Ofelt formalism is that, once the Ω_t parameters are known, the oscillator strength f can be calculated for any given (absorption or emission) transition between any pair of J states, even those not involving the ground state. This can be carried out by using expressions (6.6) and (6.7), provided that J corresponds to the departure state. For instance, this would permit us to determine the radiative rate A of an emission transition terminating in a state different to the ground state (using Equation (5.20)), as shown in Example 6.6. It should be remarked here that the radiative rate from an excited level to the ground level could be determined from an absorption spectrum, using Smakula's formula (5.22) and then Equation (5.20). However, this is not possible if the terminal state in the emission process is not populated, as it does not lead to absorption.

---

[6] (i) The excited electronic configurations $4f^{n-1}5d, \ldots$ are considered to be degenerate. (ii) There are equal energy differences between the energy states J and J' and the excited electronic configurations. (iii) All of the Stark sublevels of the ground state J are equally populated.

EXAMPLE 6.6   *The Judd–Ofelt intensity parameters experimentally obtained for the Nd:YAG laser crystal are $\Omega_2 = 0.2 \times 10^{-24}\,\text{m}^2$, $\Omega_4 = 2.7 \times 10^{-24}\,\text{m}^2$ and $\Omega_6 = 5 \times 10^{-24}\,\text{m}^2$. The most important levels for laser emissions belong to the $^4F_{3/2}$ metastable state, from which four emissions centered at about 900 nm, 1060 nm, 1340 nm, and 1900 nm are observed. (a) Determine the radiative rate A from the $^4F_{3/2}$ state. (b) If a luminescence lifetime of 230 μs is measured, estimate the quantum efficiency from this state. (Use the refractive index n = 1.82, corresponding to the main laser emission at 1060 nm.)*

(a) Using expressions (5.20), (6.6), and (6.7), we can write:

$$A_{J'J} = \frac{8\pi^2 e^2 \chi}{3\hbar\varepsilon_0 \lambda_0^3 (2J' + 1)} \times S_{J'J} \tag{6.8}$$

where λ_0 is the mean emission wavelength, $\chi = \left\{ \left[(n^2 + 2)/3 \right]^2 \times n \right\}$ includes the high-symmetry local field correction effect in Equation (5.20), and $S_{J'J} = |\mu|^2/e^2$ (note that J' denotes the departure state and J the terminal state). The $S_{J'J}$ factor (sometimes called the strength of the electric dipole transition) can be directly calculated from the Judd–Ofelt intensity parameters and the reduced matrix elements, using Equation (6.7). In our case, the emitting state is $^4F_{3/2}(J' = 3/2)$ and the terminal states corresponding to the different emissions can be easily identified from the Dieke diagram (Figure 6.1): $^4I_{9/2}$ (900 nm), $^4I_{11/2}$ (1060 nm), $^4I_{13/2}$ (1340 nm), and $^4I_{15/2}$ (1900 nm). Using the Judd–Ofelt parameters of Nd:YAG, $\Omega_2 = 0.2 \times 10^{-24}\,\text{m}^2$, $\Omega_4 = 2.7 \times 10^{-24}\,\text{m}^2$, and $\Omega_6 = 5 \times 10^{-24}\,\text{m}^2$ (Krupke, 1971), and the reduced matrix elements $|\langle \alpha J \| U^{(t)} \| \alpha' J' \rangle|$ (listed in Table 6.3) we obtain the $S_{J'J}$ factors. With this information, we can calculate the different radiative rates $A_{J'J}$ by using Equation (6.8) (remembering that it is given in MKS units). The calculated values of $S_{J'J}$ and $A_{J'J}$ are listed for the different emission transitions in Table 6.3.

Table 6.3   Values of $|\langle \alpha J \| U^{(t)} \| \alpha' J' \rangle|^2$ (reproduced with permission from Carnall *et al.*, 1968) for the transitions departing from the $^4F_{3/2}$ state. The values of $S_{J'J} = |\mu|^2/e^2$ and of the radiative probabilities $A_{J'J}$ estimated for Nd:YAG are also listed

Terminal level	$|\langle \alpha J \| U^{(2)} \| \alpha' J' \rangle|^2$	$|\langle \alpha J \| U^{(4)} \| \alpha' J' \rangle|^2$	$|\langle \alpha J \| U^{(6)} \| \alpha' J' \rangle|^2$	$S_{J'J}(\text{m}^2)$	$A_{J'J}(\text{s}^{-1})$
$^4I_{9/2}$	0	0.2296	0.0536	9×10^{-25}	1267
$^4I_{11/2}$	0	0.1423	0.4070	2.4×10^{-24}	2044
$^4I_{13/2}$	0	0	0.2117	1.05×10^{-24}	448
$^4I_{15/2}$	0	0	0.0275	1.3×10^{-25}	19

The total radiative rate from the $^4F_{3/2}$ state is, therefore, given by $A = \sum_J A_{J'J}$, where $J' = 3/2$ and $J = 9/2, 11/2, 13/2$, and $15/2$. Consequently,

$$A = 1267 + 2044 + 448 + 19 = 3778\,\text{s}^{-1}$$

which gives a radiative lifetime $\tau_0 = 1/3778 \cong 2.65 \times 10^{-4}\,\text{s} = 265\,\mu\text{s}$.

(b) Therefore, as the fluorescence lifetime is $\tau = 230\,\mu\text{s}$, the quantum efficiency is

$$\eta = \frac{230}{265} \approx 0.9$$

Certainly, this rough calculation leads to a value that is very close to the real quantum efficiency for this famous laser crystal, which is close to 1.

From the Judd–Ofelt theory, some general rules have been derived for ED transitions between $4f^n$ states of $(RE)^{3+}$ ions in crystals:

- $\Delta J \le 6$; $\Delta S = 0$, and $\Delta L \le 6$.

- For ions with an even number of electrons:

    (i) $J = 0 \leftrightarrow J' = 0$ is forbidden;

    (ii) $J = 0 \leftrightarrow$ odd J' values are weak transitions;

    (iii) $J = 0 \leftrightarrow J' = 2, 4, 6$ should be intense transitions.

These rules can be useful in establishing initial predictions, although the actual rules that govern a particular transition must be determined by group theory considerations.

6.6.2 Optical Cooling of Solids

It is very well known, and presently of paramount interest, that free atoms (in the gas phase state) can be cooled down to very low temperatures by laser illumination (*laser cooling*). The basic idea of this cooling is that, because of the Doppler effect, the atoms moving in the opposite direction to the laser beam absorb photons that have a frequency slightly lower than the resonance frequency. Due to interchange of momentum, the atoms are stopped and so they isotropically emit higher-energy photons. This is in fact an up-conversion process, and the energy difference between the absorbed and emitted photons is equal to the heat loss per emitted photon.

Optical cooling of solids is, in principle, more difficult than for gases, because the usual result of illumination is the generation of heat (phonons) and light. However, in 1995 R. I. Epstein and co-workers observed for the first time laser-induced cooling

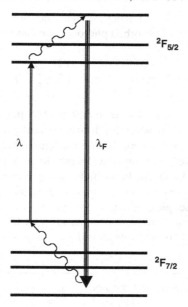

Figure 6.17 The energy-level diagram (not to scale) for the Yb^{3+} ion in ZBLANP. To make the cooling mechanism clear, the pump and emitted frequencies have been indicated by arrows and the phonon absorption processes by sinusoidal arrows (reproduced with permission from Epstein *et al.*, 1995).

of a solid. The solid was a Yb^{3+} doped heavy-metal fluoride glass (ZrF_4–BaF_2–LaF_3–AlF_3–NaF–PbF_2, abbreviated to ZBLANP). The optical cooling was made possible due to the particular energy-level structure of Yb^{3+} ions in ZBLANP, which is schematically shown in Figure 6.17. This $(RE)^{3+}$ ion involves transitions between sublevels of only two states; $^2F_{7/2}$ and $^2F_{5/2}$ (see the Dieke diagram in Figure 6.1). As a result of the crystal field acting on the Yb^{3+} ions, the ground state, $^2F_{7/2}$, splits into four Stark energy levels and the excited state, $^2F_{5/2}$, splits into three levels. The system is excited by a continuous wave laser tuned at a pump wavelength (λ) corresponding to a transition from the upper level of the ground state ($^2F_{7/2}$) to the lowest level of the excited state ($^2F_{5/2}$). Thus, the relative populations of the Stark levels in both states are changed relative to those corresponding to the situation of thermal equilibrium under no illumination. Then, the Yb^{3+} ions absorb thermal energy (phonons) from the host glass to restore the thermal equilibrium population. Therefore, the mean energy of the emitted photons (which corresponds to a wavelength λ_F) is higher than the energy of the absorbed photons ($\lambda_F < \lambda$). This anti-Stokes luminescence removes heat from the glass (cooling). Of course, successful cooling requires a high efficiency for the anti-Stokes process.

The cooling power will be equal to the emitted power minus the absorbed power, $P_{cool} = P_{em} - P_{abs}$, and it can easily be written in terms of the pump wavelength, λ.

In the ideal case of a quantum efficiency equal to one (the number of emitted photons being equal to the number of absorbed photons), we can easily obtain that

$$P_{\text{cool}}(\lambda) = P_{\text{abs}}(\lambda) \left(\frac{\lambda - \lambda_F}{\lambda_F} \right) \tag{6.9}$$

Figure 6.18 (top) shows the room temperature absorption and emission spectra of the Yb^{3+} ion in ZBLANP. The wavelength corresponding to the mean energy emitted photons, $\lambda_F = 995$ nm, has been marked on the figure. According to Equation (6.9), optical cooling only occurs for pump wavelengths larger than this wavelength. Otherwise, the system is heated. Indeed, this is made clear in Figure 6.18 (bottom), where the pump-induced temperature changes are measured by photothermal deflection spectroscopy. In this spectroscopic method, a *probe laser beam* with a wavelength that lies outside the absorption spectrum (for instance, a He–Ne laser beam) is used to illuminate the crystal sample. This *probe beam* is deflected as a result of thermally induced

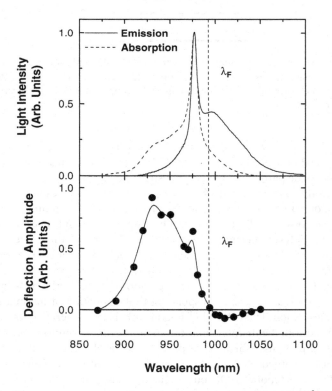

Figure 6.18   The room temperature absorption and emission spectra of Yb^{3+} ions (1 %) in ZBLANP (top). The amplitude of the photothermal-deflection waveforms as a function of the pump wavelength (bottom) (reproduced with permission from Epstein *et al.*, 1995).

gradients in the refractive index produced by the *pump absorbed beam* (i.e., by the Yb absorption–emission process). Figure 6.18 (bottom) shows the amplitude (solid circles) of the deflected He–Ne laser beam as a function of the pump wavelength, λ. It can be seen how the deflection amplitude changes its sign for pump wavelengths higher than $\lambda_F = 995$ nm. This fact reflects the fact that cooling only occurs for $\lambda > 995$ nm, while for pump wavelengths shorter than 995 nm, the Yb–ZBLANP crystal is heated.

> **EXAMPLE 6.7**   *Estimate the cooling efficiency, defined by $\eta_{\text{cool}} = \frac{P_{\text{cool}}}{P_{\text{abs}}}$, of a Yb^{3+}–ZBLANP sample when it is illuminated at room temperature by a laser tuned at 1000 nm.*
>
> From Figure 6.18, we can see that the average emission wavelength is $\lambda_F = 995$ nm. The cooling efficiency, $\eta_{\text{cool}} = P_{\text{cool}}/P_{\text{abs}}$, can easily be estimated from Equation (6.9): $\eta_{\text{cool}} = (\lambda - \lambda_F)/\lambda_F$. For pumping at $\lambda = 1000$ nm, we obtain
>
> $$\eta_{\text{cool}} = \frac{1000 - 995}{995} \approx 0.005$$
>
> Remember that this is an overestimation, in as much as in expression (6.9) we assumed an anti-Stokes quantum efficiency equal to 1.

In spite of the usually low cooling efficiencies (see the exercise above), recent experiments have demonstrated an anti-Stokes cooling from room temperature to 77 K within a certain internal volume of Yb^{3+} doped fluorochloride and fluoride glasses under high photon irradiances (Fernández *et al.*, 2000). Future practical applications of optical cooling of solids include cooling systems for spacecraft electronics and detectors, as well as for superconductive circuits.

EXERCISES

6.1 A lithium niobate crystal doped with Pr^{3+} ions is excited with light of wavelength 470 nm. Structured emission bands centered around 620 nm, 710 nm, 880 nm, and 1062 nm are observed. Using the Dieke diagram, identify the excited and terminal states responsible for the previous emission bands.

6.2 Suppose that you want to develop a phosphor based on Tb^{3+} ions that will emit in the visible region under excitation around 370 m. Using the Dieke diagram, identify the excited and terminal states related to these emissions and their related colors.

6.3 Suppose that you are going to develop an ultraviolet-emitting phosphor based on a trivalent lanthanide rare earth ion doped crystal. If you want this phosphor to present a unique emission at around 310 ± 10 nm, what trivalent rare earth ion would be the most appropriate to develop this phosphor?

6.4 The Yb^{3+} ion is commonly used to make solid state laser crystals. In which spectral range do you expect these lasers to emit ?

6.5 The fluorescence lifetime of the $^4F_{3/2}$ metastable state of Nd^{3+} ions in $LaBGeO_5$ (a solid state laser) is 280 μs and its quantum efficiency is 0.9. (a) Calculate the radiative and nonradiative rates from this excited state. (b) If the effective phonons responsible for the nonradiative rate have an energy of 1100 cm^{-1}, use the Dieke diagram to determine the number of emitted effective phonons from the $^4F_{3/2}$ excited state. (c) From which three excited states of the Nd^{3+} ions in $LaBGeO_5$ do you expect the most intense luminescence emissions to be generated?

6.6 A Gd^{3+} doped crystal is illuminated with a pulsed light source, so that the $^6I_{7/2}$ excited state of this ion is populated by absorbing 1 mJ of energy per incident pulse. Determine the heat delivered to the crystal per excitation pulse if the nonradiative rate from this state is 10^4 s^{-1}. The fluorescence lifetime of the $^6I_{7/2}$ state is 30 μs.

6.7 In Table E7.5, the fluorescence lifetimes and quantum efficiencies measured from different excited states of the Pr^{3+} ($^3P_0$ and $^1D_2$) and Nd^{3+} ($^4F_{3/2}$) ions in a $LiNbO_3$ crystal are listed. (a) Determine the multiphonon nonradiative rate from the $^4I_{9/2}$ and $^4I_{11/2}$ states of the Er^{3+} ion in $LiNbO_3$. (b) If a fluorescence lifetime of 535 μs is measured from the excited state $^2F_{5/2}$ of the Yb^{3+} ion in this crystal, estimate the radiative lifetime from this state.

Table E7.5 The lifetimes and quantum efficiencies of three excited states of trivalent rare earth ions in lithium niobate

Ion	Excited state	Lifetime (μs)	Quantum efficiency
Pr^{3+}	$^3P_0$	0.57	0.13
Pr^{3+}	$^1D_2$	89.7	1
Nd^{3+}	$^4F_{3/2}$	94.9	0.94

6.8 A certain crystal is activated with Cr^{3+} ions to develop a tunable laser. For this crystal, $Dq/B = 2$. Display the emission band (at 0 K), assuming a Huang–Rhys parameter of $S = 5$ and coupling with phonons of 200 cm^{-1}. Use the diagram in Figure 6.8 and the Racah parameter of the free ion, $B = 918$ cm^{-1}. Assume that the transition energy obtained from the Sugano–Tanabe diagram corresponds to the zero-phonon line.

6.9 Suppose that a high hydrostatic pressure is applied to the Cr^{3+} activated material of the previous exercise, so that the value of Dq/B increases up to 2.5. (a) Display the 0 K emission spectrum that you expect to occur. (b) How do you expect that this spectrum will be modified at a room temperature (300 K)?

6.10 The anion–cation distance in potassium chloride (KCl) is 0.315 nm. (a) Using the simple model of an electron in a rigid box, estimate the wavelength peaks of the two lowest energy transitions for the F centers in KCl. (b) Now determine the wavelength peak of the lowest energy transition from the experimental fit to expression (6.5) and comment on the differences relative to the result obtained in (a).

6.11 The F center in sodium fluoride (NaF) shows a broad absorption band that peaks at 335 nm (77 K). The shape of this absorption band fits a Huang–Rhys parameter of 28 and coupling with a phonon mode of 0.0369 eV. Estimate the peak position of the F center emission in NaF.

6.12 The *branching ratio* for a given emission transition, $J_i \rightarrow J_f$, is defined by $\beta = A_{J_i J_f} / \sum_k A_{J_i J_k}$, where k ranges over all the terminal levels. Using the data in Table 6.3 (in Example 6.6), determine the branching ratios from the $^4F_{3/2}$ emitting state of Nd^{3+} ions in the Nd:YAG laser crystal.

REFERENCES AND FURTHER READING

Aguilar G., M., García Solé, J., Murrieta S., H., and Rubio O., J., *Phys. Rev. B*, **26** (8), 4507 (1982).

Carnall, W. T., Fields, P. R., and Rajnak, K., *J. Chem. Phys.*, **49**, 4424 (1968).

Dawson, R., and Pooley, P., *Phys. Stat. Solidi*, **35**, 95 (1969).

Dieke, G. H., *Spectra and Energy Levels of Rare Earth Ions in Crystals*, Interscience, New York (1968).

Epstein, R. I., Buchwald, M. V., Edwards, B. C., Gosnell, T. R., and Mungan, C. E., *Nature*, **337**, October, 5000 (1995).

Fernández, J., Cussó, F., Gonzalez, R., and García, Solé, J. (eds.), *Láseres sintonizables y aplicaciones*, Ediciones de la Universidad Autónoma de Madrid, Colección de Estudios (1989).

Fernández, J., Mendioroz, A., García, A. J., Balda, R., and Adam, J. L., *Phys. Rev. B*, 5(5), 3213 (2000).

Henderson, B., and Imbusch, G. F., *Optical Spectroscopy of Inorganic Solids*, Oxford Science Publications, Oxford (1989).

Judd, B. R., *Phys. Rev.*, **127**, 750 (1962).

Kaminskii, A. A., *Crystalline Lasers: Physical Processes and Operating Schemes*, CRC Press, Boca Raton, Florida (1996).

König, E., and Kremer, s., *Ligand Field Energy Diagrams*, Plenum Press, New York (1997).

Krupke, W. F., *IEEE J. Quantum Electron.*, **QE-4**, 153 (1971).

Lifante, G., *Estudio del centro laser F_2^+*, Doctoral thesis, Universidad Autónoma de Madrid (1989).

Moulton, P. F., *Laser Focus*, **14** (May), 83 (1983).

Moulton, P. F., in *Laser Handbook*, ed. M. Bass and M. L. Stitch, North-Holland, Amsterdam (1985), p. 203.

Muñoz Santiuste, J. E., Vergara, I., and García Solé, J., *Rad. Effects Defects Solids*, **135**, 187 (1995).

Ofelt, G. S., *J. Chem. Phys.*, **37**, 511 (1962).

Powell, R. C., *Physics of Solid-State Laser Materials*, AIP Press/Springer, New York (1997).

Riseberg, L. A., and Weber, M. J., in *Progress in Optics* 14, ed. E. Wolf, North-Holland, Amsterdam (1975).

Shionoya, S., and Yen, W. M. (eds.), *Phosphor Handbook*, CRC Press, Boca Raton, Florida (1998).

Sugano, S., Tanabe, Y., and Kamimura, H., *Multiplets of Transition Metal Ions in Crystals*, Academic Press, New York (1970).

Tocho, J. O., Sanz García, J. A., Jaque, F., and García Solé, J., *J. Appl. Phys.*, **70**(10), 5582 (1991).

7

Group Theory and Spectroscopy

7.1 INTRODUCTION

Up to now, we have seen that many of the optical properties of active centers can be understood just by considering the optical ion and its local surrounding. However, even in such an approximation, the calculation of electronic energy levels and eigenfunctions is far from a simple task for the majority of centers. The calculation of transition rates and band intensities is even more complicated. Thus, in order to interpret the optical spectra of ions in crystals, a simple strategy becomes necessary.

This strategy consists of analysis of the symmetry properties for the active center (i.e., the ion and its local environment) and it is extremely useful for interpreting optical spectra without tedious calculations. Indeed, the symmetry properties of the active center are also the symmetry properties of its corresponding Hamiltonian.

Thus, among other applications, symmetry considerations are useful in solving some spectroscopic problems, such as:

- Determining the number of energy levels of a particular active center.

- Labeling these electronic energy levels in a proper way (irreducible representations) and determine their degeneracy.

- Predicting the energy level splitting induced by a reduction in symmetry (due, for instance, to pressure effects).

- Establishing selection rules for optical transitions and determining their polarization character.

An Introduction to the Optical Spectroscopy of Inorganic Solids   J. García Solé, L. E. Bausá, and D. Jaque
© 2005 John Wiley & Sons, Ltd   ISBNs: 0-470-86885-6 (HB); 0-470-86886-4 (PB)

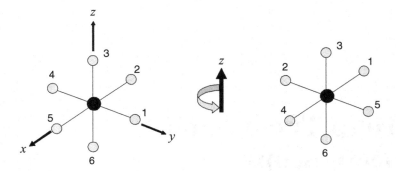

Figure 7.1   Effect of a 90° rotation around an axis of the AB_6 center.

• Determining the symmetry properties of the active center eigenfunctions.

• Analyzing the vibrations of a complex (center).

It is not the purpose of the present chapter to deal with all of the aspects related to this impressive capability. Rather, we will try to give some basic concepts, so that a nonspecialist in group theory is able to calibrate its potentiality and to apply it to simple problems in optical spectroscopy.

7.2 SYMMETRY OPERATIONS AND CLASSES

Let us consider the AB_6 center of the previous chapters, where the ligands B are now numbered from 1 to 6, as seen in Figure 7.1. If, for instance, a rotation of 90° is made around the z-axis, then the system remains the same from the physical point of view, as the B ligands 1, 2,..., 6 are indistinguishable and, of course, the Hamiltonian is unchanged. This rotation is called a C_4 (001) *symmetry operation* (where (001) denotes the rotational axis and (4) refers to the $2\pi/4$ angle rotated). This is just one among the 48 possible symmetry operations belonging to the so-called O_h *point symmetry group* for the AB_6 center. Thus, this *symmetry group* has an *order* of 48 and its elements (or symmetry operations) follow the four well-known properties of a mathematical group.[1]

At this point, it is important to mention that, in spite of the great variety of active centers (molecules, ions in solids, color centers, etc.), it can be demonstrated that only 32 point symmetry groups exist in nature. These 32 point symmetry groups (denoted by the so-called *Schöenflies symbols*) are listed in Table 7.1. The group order and

---

[1] A set of elements $G = \{a, b, c, \ldots\}$ constitutes a group if it follows the following four properties. (i) Closed by multiplication: $a \times b \in G$, for any pair of elements, a and b. (ii) Associative law: $a \times (b \times c) = (a \times b) \times c$. (iii) A unit element e exists, such that $e \times r = r \times e = r$, for any element r. (iv) For any element r, its inverse r^{-1} exists, so that $r \times r^{-1} = e$.

the number of *classes* (which will be defined below) have also been included. An inspection of Table 7.1 also shows that the 32 symmetry groups are classified into seven *crystalline systems*.

Let us continue with our illustrative AB_6 center (O_h group) in order to define a new term, the *class*, from the different symmetry operations. According to Figure 7.2(a), a clockwise rotation of $120° = 2\pi/3$ around the trigonal C_3 axis (the subscript 3 referring to the $2\pi/3$ rotation angle) modifies the ligand positions as follows:

$$1 \to 3, 3 \to 2, 2 \to 1, 4 \to 6, 6 \to 5, \quad \text{and} \quad 5 \to 4$$

Table 7.1 Point symmetry groups in crystals

Symbol	Group order	Number of classes	Crystalline system
C_1	1	1	Triclinic
C_i	2	2	Triclinic
C_s	2	2	Monoclinic
C_2	2	2	Monoclinic
C_{2h}	4	4	Monoclinic
C_{2v}	4	4	Orthorhombic
D_2	4	4	Orthorhombic
D_{2h}	8	8	Orthorhombic
C_4	4	4	Tetragonal
S_4	4	4	Tetragonal
C_{4h}	8	8	Tetragonal
C_{4v}	8	5	Tetragonal
D_{2d}	8	5	Tetragonal
D_4	8	5	Tetragonal
D_{4h}	16	10	Tetragonal
C_3	3	3	Trigonal
S_6	6	6	Trigonal
C_{3v}	6	3	Trigonal
D_3	6	3	Trigonal
D_{3d}	12	6	Trigonal
C_{3h}	6	6	Hexagonal
C_6	6	6	Hexagonal
C_{6h}	12	12	Hexagonal
D_{3h}	12	6	Hexagonal
C_{6v}	12	6	Hexagonal
D_6	12	6	Hexagonal
D_{6h}	24	12	Hexagonal
T	12	4	Cubic
T_h	24	8	Cubic
T_d	24	5	Cubic
O	24	5	Cubic
O_h	48	10	Cubic

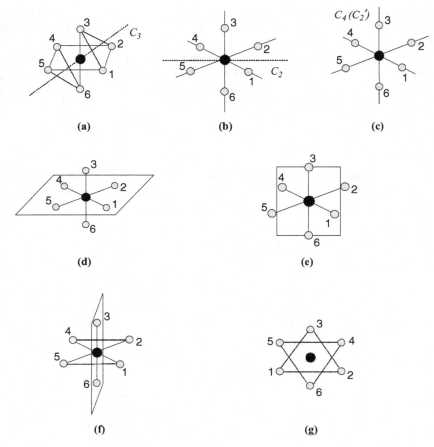

Figure 7.2   The different symmetry elements of the center AB_6. (a) A trigonal axis, C_3; (b) A binary axis, C_2. (c) A symmetry axis belonging to both the $6C_4$ and $3C_2'$ classes. (d) A symmetry plane, σ_h. (e, f) Two of the six σ_d reflection planes. (g) A view down the C_3 axis in (a) to show a roto-reflection operation, S_6.

but in a counterclockwise rotation the ligand position changes are as follows:

$$1 \to 2, 2 \to 3, 3 \to 1, 4 \to 5, 5 \to 6, \quad \text{and} \quad 6 \to 4.$$

These two symmetry operations ($2\pi/3$ clockwise and counterclockwise angle rotations) are different, but it is said that both belong to the same class. As there are four similar trigonal C_3 axes (those passing through the faces 123, 135, 354, and 234), eight operations belong to this class, which is thus designated as $8C_3$.

Let us now consider the rotations around the binary axis C_2 in Figure 7.2(b). The symmetry operations related to this axis are rotations of $2\pi/2 = 180°$. A 180°

clockwise rotation produces the following changes in the ligand positions:

$$1 \to 2, 2 \to 1, 4 \to 5, 5 \to 4, 6 \to 3, \quad \text{and} \quad 3 \to 6$$

but in the counterclockwise rotation the same changes occur, so that we write:

$$1 \leftrightarrow 2, \ 4 \leftrightarrow 5, \ 6 \leftrightarrow 3$$

where the symbol $\leftrightarrow$ denotes the lack of dependence on the rotational sense of the ligand positions. In fact, both rotations belong to the same symmetry operation. As there are six binary C_2 axes of this type, this class also contains six operations, and it is denoted by $6C_2$. Following the previous arguments and inspecting Figure 7.2(c), six operations of class C_4 are realizable, leading to the class $6C_4$. Using Figure 7.2(c) again, we achieve only three operations of class C_2', as clockwise and counterclockwise rotations are now equivalents (class $3C_2'$). Note that the axes C_2' (primed here to distinguish it from axis C_2) and C_4 are coincident.

In a similar way, we can consider other operations of the O_h point symmetry group that involve reflections. A reflection symmetry operation with respect to the plane displayed in Figure 7.2(d) leaves ligands 1, 2, 4, and 5 unchanged, while it interchanges ligands 3 and 6 ($3 \leftrightarrow 6$). As there are three of these symmetry planes (denoted by σ_h), these reflections give rise to the $3\sigma_h$ class. Figures 7.2(e) and 7.2(f) show two of the six reflection planes of another class ($6\sigma_d$) of this type.

Finally, the combination of rotations and reflections leads to a third type of symmetry operation called roto-reflection, which is usually denoted by the symbol S. An example is shown in Figure 7.2(g), which is a view down the C_3 axis of Figure 7.2(a). By means of this view, we can see that a rotation of $2\pi/6 = 60°$ around this trigonal axis locates ligand 1 above the original position of ligand 5, ligand 2 above position 6, and ligand 3 above position 4. If this rotation is followed by a reflection through a symmetry plane containing the central ion A and parallel to the plane of the page, then the following changes in ligand positions take place as a result of this combined symmetry operation:

$$1 \to 5, \ 2 \to 6, \ 3 \to 4, \ 4 \to 2, \ 5 \to 3, \quad \text{and} \quad 6 \to 1$$

It can be seen that there are seven other symmetry operations (roto-reflections) of this class, which is then denoted by $8S_6$, the subscript 6 indicating a rotation through $2\pi/6$. In a similar way, we can analyze other symmetry operations of classes $6S_4$, $1S_2$ (commonly called an inversion symmetry operation and denoted by i), and $1E$ (the identity operation that leaves the octahedron unchanged).

In summary, the O_h group contains 48 symmetry operation elements belonging to the following ten different symmetry classes:

$$E \ \ 8C_3 \ \ 6C_2 \ \ 6C_4 \ \ 3C_2' \ \ i \ \ 6S_4 \ \ 8S_6 \ \ 3\sigma_h \ \ 6\sigma_d$$

Fortunately, as we shall see in the next section, it is only necessary to work with classes rather than with symmetry operations, so that the problem is considerably simplified.

7.3 REPRESENTATIONS: THE CHARACTER TABLE

It is worthwhile now to introduce a mechanism devoted to 'representing' the symmetry operations of our AB_6 center. The symmetry operation (rotation) of Figure 7.1 transforms the coordinates (x, y, z) into $(y, -x, z)$. This transformation can be written as a matrix equation:

$$(y - x z) = (x \ y \ z) \begin{pmatrix} 0 & -1 & 0 \\ 1 & 0 & 0 \\ 0 & 0 & 1 \end{pmatrix} \tag{7.1}$$

It is then possible to represent the above-mentioned symmetry operation by the 3×3 matrix of Equation (7.1). In a more general way, we can associate a matrix **M** with each specific symmetry operation R, acting over the basic functions x, y, and z of the vector (x, y, z). Thus, we can represent the effect of the 48 symmetry operations of group O_h (AB_6 center) over the functions (x, y, z) by 48 matrices. This set of 48 matrices constitutes a *representation*, and the basic functions x, y, and z are called *basis functions*.

Obviously, we can examine the effect of the O_h symmetry operations over a different set of orthonormal basis functions, so that another set of 48 matrices (another representation) can be constructed. It is then clear that *each set of orthonormal basis functions ϕ_i generates a representation Γ* so that, as in Equation (7.1), we can write a transformation equation as follows:

$$R\phi_i = \sum_j \phi_j \Gamma^{ji}(R) \tag{7.2}$$

where R is a symmetry operation and the $\Gamma^{ji}(R)$ are the components of the matrix.

Now, if a suitable *space of basis functions* is used (a space of basis functions that is closed under the symmetry operations of the group), we can construct *a set of representations* (each one consisting of 48 matrices) for this space that is particularly useful for our purposes. [2] It is especially relevant that the matrices of each one of these representations can be made equivalent to matrices of lower dimensions.

---

[2] This set of representations is usually known as a *representation of the group*. Obviously, if we choose another *space of basis functions*, another *representation of the group* can be constructed, and so an infinite number of representations is possible for a given symmetry group.

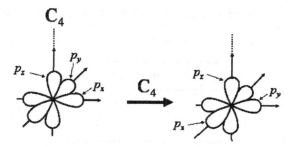

Figure 7.3   The effect on p orbitals of a $C_4(001)$ operation (clockwise sense).

The representations that involve the lowest-dimension matrices are called *irreducible representations* and have a particular relevance in group theory.

Thus, any representation Γ can be expressed as a function of its irreducible representations Γ_i. This operation is written as $\Gamma = \Sigma\, a_i \Gamma_i$, where a_i indicates the number of times that Γ_i appears in the reduction. In group theory, it is said that *the reducible representation Γ is reduced into its Γ_i irreducible representations*. The reduction operation is the key point for applying group theory in spectroscopy. To perform a reduction, we need to use the so-called *character tables*.

As an example, to construct the character table for the O_h symmetry group we could apply the symmetry operations of the AB_6 center over a particularly suitable set of basis functions: the orbital wavefunctions s, p, d, ... of atom (ion) A. These orbitals are real functions (linear combinations of the imaginary atomic functions) and the electron density probability can be spatially represented. In such a way, it is easy to understand the effect of symmetry transformations over these atomic functions.

For example, Figure 7.3 shows the effect of a $C_4(001)$ symmetry operation on the p_x, p_y, and p_z orbitals. This operation converts the orbital p_x into the orbital p_y and the orbital p_y into the orbital $-p_x$, while the orbital p_z is unaffected. We can equate these transformations as follows:

$$\left(p_y\ -p_x\ p_z\right) = \left(p_x\ p_y\ p_z\right)\begin{pmatrix} 0 & -1 & 0 \\ 1 & 0 & 0 \\ 0 & 0 & 1 \end{pmatrix} \tag{7.3}$$

so that the $C_4(001)$ operation can be represented by the square matrix of this equation.

If we were to consider the other 47 symmetry operations of the O_h group acting on the p orbitals, we would obtain a set of 48 square matrices: a representation of the O_h group associated with these orbitals. Instead of writing these matrices, let us go deeper inside the physical meaning of their traces (the sums of their diagonal elements). These traces are called *characters* in group theory. For instance, the character of the 3×3 matrix in Equation (7.3) is equal to 1. If the zero elements in this 3×3 matrix are ignored and the ± 1 elements denote the initial ($I.p.$) and final ($F.p.$) positions of the orbitals, the following scheme can represent the symmetry transformation given by Equation (7.3):

F.p. \ I.p	p_x	p_y	p_z
p_x		-1	
p_y	$+1$		
p_z			$+1$

An inspection of this scheme shows that the character $(+1)$ of the transformation matrix given in Equation (7.3) has some physical sense: in this case, it indicates that only one orbital, the p_z, is unchanged by the symmetry operation $C_4(001)$ (see Figure 7.3).

In a broader sense, the character of any symmetry transformation gives an indication of the number of orbitals that are 'unchanged' after this transformation. However, the *character can* also *be negative*, as happens with the *inversion symmetry operation i*. This operation leaves each orbital in its original position (unchanged) but with the signs of the wavefunctions reversed, as shown in Figures 7.4(a) and 7.4(b). For this case, the character is $\chi = -3$, as the three orbitals (p_x, p_y, and p_z) remain over the same axis (unchanged) but the signs of their wavefunction are reversed. Figure 7.4(c) shows the orbitals after a reflection operation σ_h through the x–y plane. The three orbitals are unchanged (they remain over the same axes) but for p_z the sign is reversed,

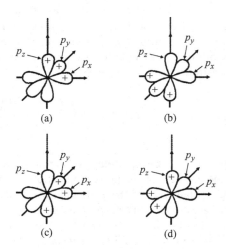

(a)   (b)

(c)   (d)

Figure 7.4   The effect of different symmetry operations over the three p orbitals: (a) the initial positions; (b) after an inversion symmetry operation; (c) after a reflection operation through the x–y plane; (d) after a rotation C_2' about the z-axis.

giving a character $\chi = 1 + 1 - 1 = 1$. The effect of a rotation C_2' about the z-axis (with a character $\chi = -1 - 1 + 1 = -1$) is shown in Figure 7.4(d).

An important conclusion envisaged from the previous paragraph is that *all of the information* needed for a symmetry operation *is contained in the character* of the matrix associated with this operation. This leads to the first great simplification: we do not need to write the full matrix associated with any transformation – its character is sufficient.

In a similar way, we can examine the effect of other symmetry operations of the O_h group over the (p_x, p_y, p_z) orbitals and so write their associated characters as follows:

E	$8C_3$	$6C_2$	$6C_4$	$3C_2'$	i	$6S_4$	$8S_6$	$3\sigma_h$	$6\sigma_d$	
3	0	−1	1	−1	−3	−1	0	1	1	(p_x, p_y, p_z)

These are the characters of the so-called T_{1u} representation of the O_h group. It should be noted that symmetry operations belonging to the same class have the same character, for a given basis function. Here, we obtain a second important simplification: it is only necessary to *work with classes, instead of invoking all of the symmetry operations* (48 in the case of the O_h group).

What has been mentioned up to now allows us to infer that the relevant information needed for a representation is given by the characters of its matrices. In fact, the full information for a given group is given by its *character table*. This table contains the character files of a particular set of representations: the *irreducible representations*. Table 7.2 shows the *character table of the O_h point group*. A character table, such as Table 7.2, contains the irreducible representations (10 for the O_h group) and their characters, the classes (also 10 for the O_h group), and the set of basis functions.

An inspection of this table is very instructive to infer some group theory properties that can be formally demonstrated:

(i) The number of *classes* is equal to the number of *irreducible representations*.

Table 7.2   A character table for the point group O_h

O_h	E	$8C_3$	$6C_2$	$6C_4$	$3C_2'$	I	$6S_4$	$8S_6$	$3\sigma_h$	$6\sigma_d$	
A_{1g}	1	1	1	1	1	1	1	1	1	1	s
A_{2g}	1	1	−1	−1	1	1	−1	1	1	−1	
E_g	2	−1	0	0	2	2	0	−1	2	0	$(d_{z^2}, d_{x^2-y^2})$
T_{1g}	3	0	−1	1	−1	3	1	0	−1	−1	
T_{2g}	3	0	1	−1	−1	3	−1	0	−1	1	(d_{xz}, d_{yz}, d_{xy})
A_{1u}	1	1	1	1	1	−1	−1	−1	−1	−1	
A_{2u}	1	1	−1	−1	1	−1	1	−1	−1	1	f_{xyz}
E_u	2	−1	0	0	2	−2	0	1	−2	0	
T_{1u}	3	0	−1	1	−1	−3	−1	0	1	1	(p_x, p_y, p_z) $(f_{x^3}, f_{y^3}, f_{z^3})$
T_{2u}	3	0	1	−1	−1	−3	1	0	1	−1	$(f_{x(y^2-z^2)}, f_{y(z^2-x^2)}, f_{z(x^2-y^2)})$

(ii) The set of characters (also called the *character*) of a representation is *unique*.

(iii) The *dimension* (that is the dimension of the matrices) of each irreducible representation *is given by the character corresponding to class E.*

In the so-called *Mulliken notation*, representations A and B (which does not appear in Table 7.2) are mono-dimensional, E representations are bi-dimensional, and T representations are three-dimensional. Other irreducible representations of higher order are G (three-dimensional) and H (tetra-dimensional). We will also state now that the dimension of a representation gives the degeneracy of its associated energy level.

In the Mulliken notation, the subscripts u *(ungerade* = odd) and g *(gerade* = even) indicate whether an irreducible representation is symmetric (g) or anti-symmetric(u), in respect to the inversion operation (i).

Another common notation for the irreducible representations is the so-called *Bethe notation*, in which the representations are denoted by Γ_i symbols (i = 1, 2, ...), where the subscript i denotes the dimension. It is not simple to establish an equivalence between these two types of notation, since it depends on the symmetry group. For the moment, we will just mention both notations so that readers will be familiar with any character table.

7.4 REDUCTION IN SYMMETRY AND THE SPLITTING OF ENERGY LEVELS

Once we have verified the utility of the information contained in a character table, the next step is to start to use character tables in spectroscopy. Let us now consider a simple case, in which the symmetry of an optical center (e.g., the AB_6 center) is reduced in some way; for instance, by applying an axial pressure. We know that, in general, this symmetry reduction is accompanied by a splitting of energy levels. Group theory is very useful to unequivocally predict the number of split components and their degeneracy (of course, the amount of splitting for a given energy level cannot be predicted by group theory).

To tackle this problem, we first need to know how a given representation Γ is reduced to its irreducible representations Γ_i; in other words, to determine the coefficients a_i in the equation $\Gamma = \Sigma\, a_i\, \Gamma_i$. Although this is a key problem in group theory, here we only explain how to perform this reduction without entering into formal details, which can easily be found in specialized textbooks.

It can be shown that the number of times a_i that the irreducible representation Γ_i appears in Γ is given by

$$a_i = \frac{1}{g} \sum n_R \chi(R)\chi_{\Gamma_i}(R) \tag{7.4}$$

where the sum is extended over all of the classes, g is the group order, $\chi(R)$ and $\chi_{\Gamma_i}(R)$ are, respectively, the characters of Γ and Γ_i for the operation R, and n_R is the number of operations in the class to which R belongs. Example 7.1 is an instructive case of how to apply Equation (7.4) in a reduction symmetry problem.

EXAMPLE 7.1 *Reduction of a representation.*

Let us consider the character table of the C_{4v} group and the representation Γ displayed below (see Table 7.3). Γ is reducible, because its character does not coincide with any one of the irreducible representations of the C_{4v} group. According to Equation (7.4) and the character table (Table 7.3), we can write

$$a_{A_1} = \tfrac{1}{8}\left[1 \times 6 \times 1 + 2 \times 0 \times 1 + 1 \times 2 \times 1 + 2 \times 0 \times 1 + 2 \times 0 \times 1\right] = 1$$

so that we say that the irreducible representation A_1 appears once in the reduction of Γ. In a similar way, we obtain

$$a_{A_2} = 1, \quad a_{B_1} = 1, \quad a_{B_2} = 1, \quad \text{and} \quad a_E = 1$$

Consequently, the other irreducible representations, A_2, B_1, B_2, and E, also appear only once. Therefore, the reduction equation can be written as follows:

$$\Gamma = A_1 + A_2 + B_1 + B_2 + E$$

The same result can sometimes be easily inferred from direct inspection of the character table. In fact, we can directly observe from Table 7.3 that $\chi^{A_1} + \chi^{A_2} + \chi^{B_1} + \chi^{B_2} + \chi^E = \chi^\Gamma$. Because there is only one possible decomposition, we conclude that the representation Γ is decomposed into the irreducible representations A_1, A_2, B_1, B_2, and E of group C_{4v}, and that each irreducible representation appears only once in the reduction.

Table 7.3 The character table of group C_{4v}. The basis functions are not included for the sake of brevity. A reducible representation, Γ, is shown below

C_{4v}	E	$2C_4$	C_2	$2\sigma_v$	$2\sigma_d$
A_1	1	1	1	1	1
A_2	1	1	1	-1	-1
B_1	1	-1	1	1	-1
B_2	1	-1	1	-1	1
E	2	0	-2	0	0
Γ	6	0	2	0	0

We will now connect with spectroscopic problems, to envisage the utility of what we have seen up to now.

Let us consider an energy level of an active center with energy E_n and degeneracy d_n. This level is associated with a set of eigenfunctions $\{\phi_1, \ldots, \phi_i, \ldots, \phi_{d_n}\}$ that belongs to a space of eigenfunctions (*basis functions*) of the center. If R is a symmetry operation, then $R\phi_i$ is a new function, which represents the effect of R on ϕ_i leading to an equivalent system. Therefore, the function $R\phi_i$ must be one of the eigenfunctions associated with that same energy level E_n, and so it must belong to the set of functions $\{\phi_1, \ldots, \phi_i, \ldots, \phi_{d_n}\}$. This means that *this set of functions generates a representation* Γ_n (i.e., a set of matrices, each one related to a symmetry operation R) of the symmetry group G of the active center. We postulate that this representation must be irreducible,[3] so that it is established that:

> The eigenfunctions of the Hamiltonian H related to an energy level E_n belong to one of the irreducible representations Γ_n of the symmetry group G, and the eigenfunctions belonging to another irreducible representations Γ_m of G are related to another energy level E_m.

This important postulate indicates that *we can label the energy levels of an active center by irreducible representations* of its symmetry group G. The group G is usually called the *Hamiltonian symmetry group.* [4]

We can check the validity of the previous assessment by considering the splitting of a d^1 electron in a point symmetry group O_h. In Section 5.2, we demonstrated that the outer d^1 electronic configuration of an active center splits into two energy levels, separated by an amount of energy known as $10Dq$. These energy levels are associated with the irreducible representations E_g and T_{2g} of the O_h group. The split into these two components can be easily inferred from the character table of group O_h, given in Table 7.2. From this table, we can observe that the orbitals (d_{xy}, d_{yz}, d_{xz}) transform as the T_{2g} representation, while the orbitals $(d_{z^2}, d_{x^2-y^2})$ transform as another irreducible representation, the E_g representation.

At this point, we are ready to treat several spectroscopic properties that are related to a reduction in symmetry of the active center. We will now give two relevant examples (7.1 and 7.2) devoted to learning how group theory can be successfully applied to the determination of the split components related to a reduction in the local symmetry of a center. In the first example (7.2), this symmetry reduction is induced by applying an axial pressure to a Cr^{3+} ion doped crystal. The second example (7.3) is given to learn the steps needed for labeling the energy levels of an active center in a crystal. In fact, this is also a symmetry reduction problem, as it involves passing from full symmetry (the free ion) to a specific local symmetry (an ion in a crystal).

---

[3] Otherwise, it would mean that there are at least two subspaces of the space of eigenfunctions $\{\phi_1, \ldots, \phi_i, \ldots, \phi_{gn}\}$, each of them closed under the symmetry operations of G. This would mean that there are no symmetry operations connecting these two subspaces, in spite of the fact that they have the same energy E_n. This, of course, seems to be unreasonable except in the case of accidental coincidence of two energy levels.

[4] This definition is because there are cases in which the Hamiltonian symmetry group has more elements (double) than the point symmetry group of the active center. Those cases deal with rare earth ions with half-integer J values; for instance, the Nd^{3+} ion. They will be treated in Section 7.7.

EXAMPLE 7.2 *The effect of an axial pressure in a MgO:Cr^{3+}crystal.*

The Cr^{3+} ion has an outer electronic configuration d^3. In the MgO crystal, each Cr^{3+} ion is surrounded by six O^{2-} ligand ions in octahedral symmetry, as depicted in Figure 7.5(a). The actual symmetry of this center corresponds to the O Hamiltonian symmetry group (a subgroup of O_h).[5] From the character table of the O group (see Table 7.4) we realize that the energy levels of Cr^{3+} in MgO can be labeled by representations A_1, A_2, E, T_1, or T_2 (the irreducible representations of group O).

By applying a pressure along the [001] axis (using a diamond cell), the local symmetry around the Cr^{3+} ions is reduced and energy level splitting can occur. The new distorted environment, induced by the displacement of two O^{2-} ions toward the Cr^{3+} ion (see Figure 7.5(b)), corresponds to a D_4 symmetry (a subgroup of group O). We can use group theory to predict the number of split components due to this symmetry reduction, and label them using irreducible representations of the D_4 group.

The irreducible representations of Cr^{3+} in MgO (the O group) are now reducible in the lower-symmetry D$_4$ group. Thus, the first step is to rewrite the representations A_1, A_2, E, T_1, and T_2 (the irreducible representations in O)

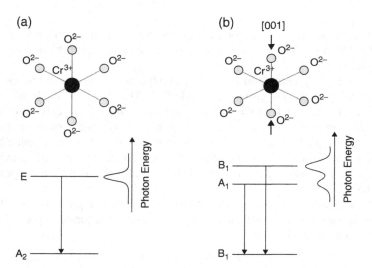

Figure 7.5 Symmetry reduction by an axial external pressure in a MgO:Cr^{3+} crystal and its effect on the red emission and energy levels of Cr^{3+} in MgO: (a) the undistorted center (O symmetry) and (b) the distorted center (D_4 symmetry), after the axial pressure is applied.

---

[5] This is because the Cr^{3+} ions are not exactly on the centers of the octahedrons and so they have not been subjected to a local inversion symmetry.

Table 7.4 The character tables of group O and its subgroup D_4. The irreducible representation T_1 of group O appears written below as a reducible representation in D_4

O	E	$8C_3$	$3C_2$	$6C_2'$	$6C_4$	
A_1	1	1	1	1	1	$x^2 + y^2 + z^2$
A_2	1	1	1	-1	-1	xyz
E	2	-1	2	0	0	$(x^2 - y^2, 3z^2 - r^2)$
T_1	3	0	-1	-1	1	$(x, y, z)(R_x, R_y, R_z)$
T_2	3	0	-1	1	-1	(yz, zx, xy)

D_4	E	$C_2{}^z$	$2C_2{}^{x,y}$	$2C_2'$	$2C_4$	
A_1	1	1	1	1	1	$(x^2 + y^2); z^2$
A_2	1	1	-1	-1	1	$z; R_z$
B_1	1	1	1	-1	-1	$x^2 - y^2$
B_2	1	1	-1	1	-1	xy
E	2	-2	0	0	0	$(x, y)(xz, yz)(R_x, R_y)$

T_1	3	-1	-1	-1	1	

in the character table of group D_4. These representations will now become reducible in D_4. The second step is to solve the reduction problem, as given in Example 7.1.

The T_1 representation of the O group has been rewritten below the character table of the D_4 group (see Table 7.4). This has been made taking the character values of T_1 in the O group corresponding to those classes related to the D_4 subgroup; that is, classes with coincident symmetry elements in both groups. It can be seen how this representation $\Gamma = T_1$ is now reducible in group D_4. A direct inspection of the D_4 character table (Table 7.4) shows that $T_1 = A_2 + E$, although this result can be formally obtained as in the previous example.

In a similar way, we can write the other irreducible representations of O as representations of D_4 and then decompose them into irreducible representations of this group. We obtain:

$$T_2 = B_2 + E, \quad E = B_1 + A_1, \quad A_2 = B_1 \quad \text{and} \quad A_1 = A_1$$

Figure 7.5(b) shows a nice experimental confirmation of a splitting predicted by group theory. The red emission of Cr^{3+} in MgO, related to the E $\rightarrow$ A$_2$ transition, splits into two emissions as a result of an applied pressure (in the order of 10 kg mm^{-2}). The presence of this double emission is predicted by group theory as the excited E level (the irreducible representation in group O) splits into B_1 and A_1 levels (irreducible representations in group D_4).

The previous example has shown how group theory can be used in a symmetry reduction problem. This symmetry reduction also occurs when an ion is incorporated in a crystal. We will now treat how to predict the number of energy levels of the ion in the crystal (the active center) and how to properly label these levels by irreducible representations.

EXAMPLE 7.3   *The construction of representations: the full rotation group.*

A very common problem in spectroscopy is to predict and label the energy levels of an active ion in a crystal from the energy levels of the free ion; that is, from the energy levels of the ion out of the crystal. This 'free ion' case can be treated like that corresponding to a group that contains all rotational angles about any axis. This group has an infinite number of symmetry elements and is known as the *full rotation group*. In this group, all rotations through the same angle about any axis belong to the same class. Therefore, there is an infinite number of classes and, consequently, an infinite number of irreducible representations.

Considering that the free ion energy-level scheme can be described by $^{2S+1}L_J$ states, the procedure is to construct the irreducible representations in the full rotation group corresponding to these states. These representations will be labeled as D^J and, in general, they correspond to reducible representations in the symmetry group of the ion in the crystal. Let us call this generic group G. Once the D^J representations have been constructed, the next step is to carry out the decomposition of these representations into irreducible representations Γ_i of the G group; that is, $D^J = \Sigma \, a_i \Gamma_i$.

We know that each free ion level D^J is associated with $2J + 1$ functions whose angular parts have the form $e^{i m_J \phi}$, where m_J is a quantum number that takes $2J + 1$ values ($m_J = J, J - 1, \ldots, -J$) and the angle ϕ is a polar coordinate, as shown in Figure 7.6. The angular part of the wavefunction is what distinguishes the various orbitals. A rotation through α around the z-axis, C_α (as in Figure 7.6), modifies this angular part and can be expressed by the equation:

$$C_\alpha \left(e^{iJ\phi}, e^{i(J-1)\phi}, \ldots, e^{-iJ\phi} \right) = \left(e^{iJ(\phi+\alpha)}, e^{i(J-1)(\phi+\alpha)}, \ldots, e^{-iJ(\phi+\alpha)} \right) \quad (7.5)$$

This rotation, C_α, can be also equated by using the corresponding matrix:

$$\begin{pmatrix} e^{iJ\alpha} & 0 & . & . & 0 \\ 0 & e^{i(J-1)\alpha} & . & . & . \\ . & . & . & . & . \\ . & . & . & . & 0 \\ 0 & . & . & . & e^{-iJ\alpha} \end{pmatrix} \quad (7.6)$$

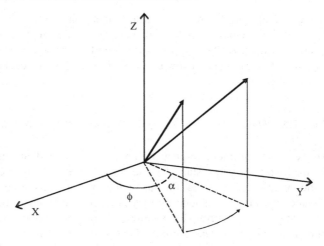

Figure 7.6 The change in the polar coordinate ϕ of an electron due to rotation through an angle α around the z-axis.

The character of this operation is the trace of this matrix, $\chi(\alpha) = \sum\limits_{m=-J}^{m=+J} e^{im\alpha}$. This expression corresponds to a summation over a geometric series, so that

$$\chi(\alpha) = \frac{\sin(J + 1/2)\alpha}{\sin(\alpha/2)} \qquad (7.7)$$

Equation (7.7) can be used to determine the characters of the symmetry operations C_n, where $n = 2\pi/\alpha$, and then, to construct the reducible representations D^J in the group G of the ion in the crystal. Some of the most common character elements are listed below:

$$
\begin{aligned}
\chi^J(E) &= 2J + 1 \\
\chi^J(\pi) &= (-1)^J \\
\chi^J(\pi/2) &= 1 \qquad \text{for } J = 0, 1, 4, 5, 8, 9, \dots \\
&= -1 \qquad \text{for } J = 2, 3, 6, 7, \dots \\
\chi^J(2\pi/3) &= 1 \qquad \text{for } J = 0, 3, 6, \dots \\
&= 0 \qquad \text{for } J = 1, 4, 7, \dots \\
&= -1 \qquad \text{for } J = 2, 5, 8, \dots
\end{aligned}
\qquad (7.8)
$$

At this point, we are able to construct the reducible representations D^J of a group composed only of rotational elements. For instance, let us consider that the ion in the crystal has a symmetry group $G = O$, whose character table (Table 7.4) consists of only rotational symmetry elements: classes C_n.

Table 7.5   The character of the representations D^J. The decomposition of the representations D^J into irreducible representations of O group

D^J representation	O classes E $(\alpha = 2\pi)$	$8C_3$ $(\alpha = 2\pi/3)$	$3C_2$ $(\alpha = \pi)$	$6C_2'$ $(\alpha = \pi)$	$6C_4$ $(\alpha = \pi/2)$	Irreducible representation in O
D^0	1	1	1	1	1	A_1
D^1	3	0	-1	-1	1	T_1
D^2	5	-1	1	1	-1	$E + T_2$
D^3	7	1	-1	-1	-1	$A_2 + T_1 + T_2$
D^4	9	0	1	1	1	$A_1 + E + T_1 + T_2$
D^5	11	-1	-1	-1	1	$E + 2T_1 + T_2$
D^6	13	1	1	1	-1	$A_1 + A_2 + E + T_1$ $+ 2T_2$

In Table 7.5, we show the *character* (defined as the set of character elements of a representation) of different representations D^J (from $J = 1$ to 6) of the O group. The character elements were obtained from Equation (7.7). These D^J representations, which were irreducible in the full rotation group, are in general reducible in O, as can be seen by inspecting the character table of O (in Table 7.4). Thus, the next step is to decompose them into irreducible representations of O, as we did in Example 7.1. Table 7.5 also includes this reduction; in other words, the irreducible representations of group O into which each D^J representation is decomposed. We will use this table when treating relevant examples in Section 7.6.

7.5 SELECTION RULES FOR OPTICAL TRANSITIONS

Group theory can also be applied to determine whether an optical transition is allowed in a particular optical center. As we showed in Section 5.3, the probability of a radiative transition between two given states, ψ_i (initial) and ψ_f (final), is proportional to

$$| < \psi_f | \boldsymbol{\mu} | \psi_i > |^2 \tag{7.9}$$

where $\boldsymbol{\mu} = \sum_i e\mathbf{r}_i$ is the electric dipole moment operator, for electric dipole transitions, and $\boldsymbol{\mu} = \sum_i (e/2m)(\mathbf{l}_i + 2\mathbf{s}_i)$ is the magnetic moment operator, for magnetic dipole transitions.

Thus, the band intensity (absorption or emission) related to an optical transition between two states ψ_i and ψ_f depends on the value of the matrix element given in

Equation (7.9). By analyzing this matrix element, we can establish the selection rules for the transition.

Obviously, the symmetry properties of the center affect the value of the matrix element given in Equation (7.9). We know from Section 7.4 that the states ψ_i and ψ_f belong to different irreducible representations, Γ_i and Γ_f, in the group symmetry of the center. In fact, these irreducible representations are used to label the energy levels associated with these states. The electric dipole moment operator μ corresponds to a function whose symmetry is related to an irreducible representation, Γ_μ, of the symmetry group center. For instance, the electric dipole operator $e(x, y, z)$ has the same symmetry as the (p_x, p_y, p_z) orbitals, and then it transforms in the same way for a particular point group. Thus, if we consider the O_h group, we see from Table 7.2 that the electric dipole operator transforms as the irreducible representation T_{1u}. However, as we see from Table 7.4, for group O the electric dipole operator transforms as T_1, the representation corresponding to the (x, y, z) functions.

At this point, we should invoke the so-called Wigner–Eckart theorem, whose demonstration is beyond the scope of this book (see, e.g., Tsukerblat, 1994). From this theorem, it is possible to establish the following selection rule:

The matrix element $| < \psi_f \, | \, \mu \, | \, \psi_i > |^2$ *[Equation (7.9)] is zero unless the direct product between the irreducible representations* Γ_i *and* Γ_μ *contains the irreducible representation associated with the final state* Γ_f.

This selection rule can be expressed as:[6]

$$\Gamma_i \times \Gamma_\mu \subset \Gamma_f \tag{7.10}$$

Here we find a new concept, the *direct product* between irreducible representations of a symmetry group. This direct product is related to the product of their corresponding space functions. For our purposes, we will only mention that the direct product between two, Γ_j and Γ_k, (or more) irreducible representations of a group is a new representation whose character is given by $\chi^{\Gamma_j \times \Gamma_k}(R) = \chi^{\Gamma_j}(R) \times \chi^{\Gamma_k}(R)$. This *representation product* is, in general, reducible, although the original representations were irreducible. Thus, the direct product $\Gamma_i \times \Gamma_\mu$ of Equation (7.10) will be a new representation Γ_p that, in general, will be reducible. If the irreducible representation Γ_f appears after reducing Γ_p, then the transition $\Gamma_i \rightarrow \Gamma_f$ is allowed; otherwise, it is forbidden.

The next example is provided to clarify how to work with direct products between irreducible representations. Below, in Section 7.6, we will apply the selection rule given by Equation (7.10) to relevant examples.

---

[6] It can be shown that the selection rule given by Equation (7.10) is equivalent to $\Gamma_i \times \Gamma_\mu \times \Gamma_f \subset \Gamma_1$, where $\Gamma_1(A_1$ in the Mulliken notation) is the identity representation.

EXAMPLE 7.4   *The direct product between irreducible representations of group O.*

If we consider the sets of functions (x_1, y_1, z_1) and (x_2, y_2, z_2), both of which belong to the same representation T_1 of the O group, the *product functions* constitute a nine-dimensional space. Consequently, these product functions belong to a nine-dimensional representation, denoted by $T_1 \times T_1$. It can be shown that the character of this representation is given by $\chi^{T_1 \times T_1}(R) = \chi^{T_1}(R) \times \chi^{T_1}(R)$. This character can easily be obtained from the character table of the O group (Table 7.4):

$$\chi^{T_1 \times T_1} = 9 \ 0 \ 1 \ 1 \ 1$$

We see that this representation is reducible, so it can be decomposed into its irreducible representations in O, giving:

$$T_1 \times T_1 = A_1 + E + T_1 + T_2$$

Similarly, we could perform other direct products between irreducible representations of the O group, and then decompose them into irreducible representations of this group. We would obtain:

$$A_1 \times A_1 = A_1; A_1 \times A_2 = A_2; A_1 \times E = E; A_1 \times T_1 = T_1; A_1 \times T_2 = T_2$$

$$A_2 \times A_2 = A_1; A_2 \times E = E; A_2 \times T_1 = T_2; A_2 \times T_2 = T_1$$

$$E \times E = A_1 + A_2 + E; E \times T_1 = T_1 + T_2; E \times T_2 = T_1 + T_2$$

$$T_1 \times T_2 = A_2 + E + T_1 + T_2$$

$$T_2 \times T_2 = A_1 + E + T_1 + T_2$$

Fortunately, this information is generally contained in the so-called *multiplication table*s of each point symmetry group. These tables are available in specific group theory textbooks.

7.6 ILLUSTRATIVE EXAMPLES

At this point in the chapter, we know how to apply group theory to solve the problem of labeling the energy levels of an optical center as well as to determine which transitions are optically allowed. The next few examples are devoted to practice with these aspects.

EXAMPLE 7.5   *Emissions of Eu^{3+} ions in octahedral symmetry (O group).*

The Eu^{3+} ion is commonly used as an activator in many red-emitting phosphors. Its red emission is due to transitions from the $^5D_0$ excited state (see the Dieke's diagram in Figure 6.1). Thus, the possible transitions are those terminating in the ^7F_J ($J = 0, 1, 2$) states (see Figure 7.7). By group theory we can predict the possible number of emission transitions (i.e., the number of energy levels involved) and whether they obey the electric dipole or magnetic dipole selection rules.

The first step is to construct the D^J representations in group O. Using expressions (7.7) and (7.8) for $J = 0, 1, 2$, we can determine the characters for these representations in O. These characters have already been given in Table 7.5. The next step is to decompose each representation D^J into its irreducible representations in O. This is a problem similar to that of Example 7.1. It can be checked that:

$$D^0 = A_1, \qquad D^1 = T_1, \qquad D^2 = E + T_2$$

Thus, the expected energy-level scheme for the Eu^{3+} ions in the crystal (O symmetry) is that displayed in Figure 7.7. However, it should be recalled here that by group theory we cannot know neither the energy location of each level nor the energy order of these levels.

Let us now apply group theory again to determine the electric dipole (ED) allowed transitions.

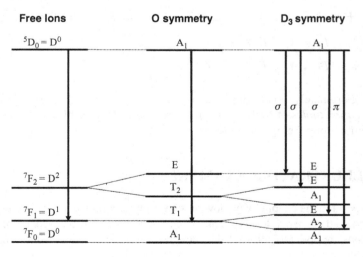

Figure 7.7   The energy-level scheme and allowed ($\rightarrow$) ED transitions of free Eu^{3+} ions and Eu^{3+} ions in crystals (O and D_3 symmetry local environments).

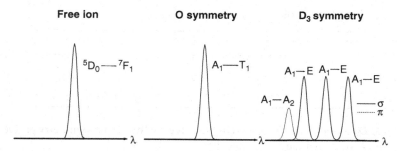

Figure 7.8   The scheme of the lines expected in the emission spectrum (at the electric order dipole) of free Eu^{3+} ions and Eu^{3+} ions in the O and D_3 local environments.

In the ideal case of free Eu^{3+} ions, we first must observe that the components of the electric dipole moment, $e(x, y, z)$, belong to the irreducible representation D^1 in the full rotation group. This can be seen, for instance, from the character table of group O (Table 7.4), where the dipole moment operator transforms as the T_1 representation, which corresponds to D^1 in the full rotation group (Table 7.5). Since $D^0 \times D^1 = D^1$, only the $^5D_0 \rightarrow \, ^7F_1$ transition would be allowed at electric dipole order. This is, of course, the well known selection rule $\Delta J = 0, \pm 1$ (except for $J = 0 \rightarrow J = 0$) from quantum mechanics. Thus, the emission spectrum of free Eu^{3+} ions would consist of a single $^5D_0 \rightarrow \, ^7F_1$ transition, as indicated by an arrow in Figure 7.7 and sketched in Figure 7.8.

In the case of Eu^{3+} ions embedded in the crystal (O symmetry group) we could, in principle, expect the following four emission lines: $A_1(^5D_0) \rightarrow A_1(^7F_0), T_1(^7F_1), T_2(^7F_2)$, and $E(^7F_2)$, where the free ion states from which each level originates have been indicated in brackets. The electric dipole moment operator transforms as the functions (x, y, z) and so it belongs (see Table 7.4) to the irreducible representation T_1. Therefore, for optical transitions departing from the A_1 excited level, the selection rule is $A_1 \times T_1 = T_1$, so that only the $A_1(^5D_0) \rightarrow T_1(^7F_1)$ emission is allowed in the ED approximation, as displayed in Figure 7.7. Consequently, we should also observe a single emission line, as sketched in Figure 7.8.

EXAMPLE 7.6   *Emissions of Eu^{3+} ions in a D_3 group symmetry: polarized transitions.*

In many crystals, the local environment of Eu^{3+} ions is lower than the octahedral O symmetry and this can lead to polarized transitions.

Let us suppose a D_3 local symmetry for the Eu^{3+} ions in a particular crystal. Following the same procedure as in Example 7.2, and using the character table

Table 7.6   The character table of group D_3

D_3	E	$2C_3$	$3C_2'$	
A_1	1	1	1	$(x^2 + y^2); z^2$
A_2	1	1	-1	z
E	2	-1	0	$(x, y); (xz, yz)\ (x^2 - y^2, xy)$

of the D_3 group (Table 7.6), it is easy to solve the 'reduction symmetry problem' from O to D_3 symmetry to show that

$$A_1 = A_1, \qquad E = E, \qquad T_1 = A_2 + E, \qquad T_2 = A_1 + E$$

Consequently, we expect that the T_2 and T_1 levels of the Eu^{3+} ion (in O symmetry) will each split into two levels in the lower D_3 symmetry, as shown in Figure 7.7.

Let us now determine the ED allowed transitions departing from the A_1 excited state. According to the character table of the D_3 group (see Table 7.6), the (x, y) function belongs to the E irreducible representation, while the z function belongs to the A_2 irreducible representation. This means that, in this symmetry group, the ED selection rules are different depending on the polarization of the light.

Thus, the electric dipole σ polarized emissions departing from the A_1 level (in which the electric field of the emitted light is parallel to x or y) are those defined by the direct product $A_1 \times E$. An inspection of Table 7.6 shows that $A_1 \times E = E$, so that only $A_1 \to E$ emissions are allowed for σ emitted radiation (as shown in Figures 7.7 and 7.8).

Electric dipole π polarized emissions are those in which the electric field of the emitted light is parallel to z. Thus the selection rule for π emissions from level A_1 are those defined by the direct product $A_1 \times A_2$. By an inspection of the character table of group D_3 (Table 7.6), we can easily prove that $A_1 \times A_2 = A_2$, so that only the $A_1 \to A_2$ emission is allowed by π polarized radiation (as shown in Figures 7.7 and 7.8).

It is interesting to observe that while for free ion or O symmetries only one ED emission is expected to occur, a symmetry reduction to D_3 induces the appearance of four emission lines. Three of these emissions are σ polarized and one is π polarized, as shown in Figure 7.8.

7.7   ADVANCED TOPIC: THE APPLICATION TO OPTICAL TRANSITIONS OF KRAMERS IONS

So far, we have seen how to use the character tables of the point symmetry groups to interpret the optical spectra of some ions. When dealing with spin functions of ions, we have the possibility of half-odd integer values for the spin (or for the spin-plus-orbital) functions. In order to apply group theory to these ions, we need to extend our concept of group symmetry to the so-called *double symmetry groups*. These groups, and their use in spectroscopy, are introduced in the next example regarding trivalent rare earth ions with half-odd integer J values. These ions are known as *Kramers ions* as they follow the Kramers theorem, which establishes that all electronic levels of atoms (ions) with an odd number of electrons must be at least doubly degenerate in the absence of a magnetic field.

EXAMPLE 7.7   *The luminescence of Sm^{3+} in the $YAl_3(BO_3)_4$ crystal.*

Yttrium aluminum borate, $YAl_3(BO_3)_4$ (abbreviated to YAB), is a nonlinear crystal that is very attractive for laser applications when doped with rare earth ions (Jaque *et al.*, 2003). Figure 7.9 shows the low-temperature emission spectrum of Sm^{3+} ions in this crystal. The use of the Dieke diagram (see Figure 6.1) allows to assign this spectrum to the $^4G_{5/2} \rightarrow {}^6H_{9/2}$ transitions. The polarization character of these emission bands, which can be clearly appreciated in Figure 7.9, is related to the D_3 local symmetry of the Y^{3+} lattice ions, in which the Sm^{3+} ions are incorporated. The purpose of this example is to use group theory in order to determine the Stark energy-level structure responsible for this spectrum.

The $4f^5$ outer electronic configuration of Sm^{3+} ion leads to states with half-odd integer values of J (see Figure 6.1), due to the half-odd integer values for the total spin S. In these cases, there are some peculiarities in the rotations of

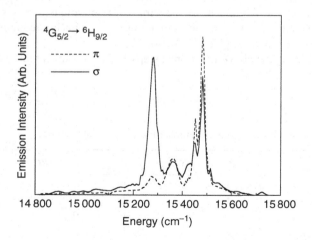

Figure 7.9 The emission spectrum (at 10 K) of Sm^{3+} ions in YAB (reproduced with permission from Cavalli *et al.*, 2003).

the spin functions. Using Equation (7.7) we find that, for a rotation of $\alpha + 2\pi$,

$$\chi^J(\alpha + 2\pi) = (-1)^{2J}\chi^J(\alpha) \tag{7.11}$$

Then, for half-odd integer values of J, a rotation through an angle of $\alpha + 2\pi$ reverses the sign of the character corresponding to a rotation through α. However, a rotation through $\alpha + 4\pi$, leads to the same character as a rotation through α. This peculiar behavior is due to the spin functions and it produces, in principle, a double-value character for each rotation. Therefore, in order to apply group theory properly, the so-called *double groups* must be now considered.

Table 7.7 shows the character table for the $\overline{D}_3$ double group (the upper line is used to denote such kinds of group), which can be compared with Table 7.6 for the single D_3 group. To simplify the notation of the classes, an additional rotation through 2π is written as R. It can be seen that the number of classes (6) for the $\overline{D}_3$ double group is twice that for the single group D_3 (3). Other peculiarities are an even dimension for the new representations ($E_{1/2}$ and $E_{3/2}$) and the appearance of complex characters.

Once these *double-group character tables* are known, the procedure is exactly the same as that followed in the preceding sections when single character tables were used. Let us now tackle the problem of understanding the spectrum given in Figure 7.9 for Sm^{3+} ions in YAB.

The first step is to construct the representations $D^{5/2}$ and $D^{9/2}$, corresponding to the excited $^4G_{5/2}$ and terminal $^6H_{9/2}$ states, in the $\overline{D}_3$ double group. This can be done as in Example 7.3, but using Equation (7.7) together with Equation (7.11), which gives the characters for double groups. As both representations ($D^{5/2}$ and $D^{9/2}$) will appear as reducible representations, the next step is to reduce them into irreducible representations of $\overline{D}_3$. This is a reduction problem similar to the examples given in Section 7.4.

Following the above-mentioned steps, it can be shown that

$$D^{5/2} = 2E_{1/2} + E_{3/2} \qquad \text{and} \qquad D^{9/2} = 3E_{1/2} + 2E_{3/2}$$

Table 7.7 The character table for the $\overline{D}_3$ double group

$\overline{D}_3$	E	R	C_3, RC_3^2	C_3^2, RC_3	$3C_2'$	$3RC_2'$	
A_1	1	1	1	1	1	1	
A_2	1	1	1	1	-1	-1	z
E	2	2	-1	-1	0	0	x, y
$E_{1/2}\begin{cases} \\ \end{cases}$	1	-1	-1	1	i	$-i$	
	1	-1	-1	1	$-i$	i	
$E_{3/2}$	2	-2	1	-1	0	0	

Thus, the excited state $^4G_{5/2}$ (or $D^{5/2}$) splits into three Stark levels in $\overline{D}_3$ symmetry, labeled by $E_{1/2}$, $E_{1/2}$, and $E_{3/2}$, while the terminal state $^6H_{9/2}$ (or $D^{9/2}$) splits into five Stark levels, $E_{1/2}$, $E_{1/2}$, $E_{1/2}$, $E_{3/2}$, and $E_{3/2}$. In fact, the five peaks observed in the emission spectrum of Figure 7.9 are related to these five terminal levels. This is because, at the low temperature (10 K) of the spectrum, radiative transitions only depart from the lowest level of the excited state.

Now, to make a proper assignment, we establish the ED selection rules between the possible energy levels ($E_{1/2}$ or $E_{3/2}$), following the same procedure as in Example 7.6 but using the *double-group character table* of $\overline{D}_3$ (Table 7.7). According to this character table, the p_z electric dipole moment component, $p_z = ez$, belongs to the A_2 irreducible representation. Thus the electric dipole π ($E \parallel z$) polarized transitions from $E_{1/2}$ or $E_{3/2}$ energy levels are related to the following direct products:

$$E_{1/2} \times A_2 = E_{1/2} \quad \text{and} \quad E_{3/2} \times A_2 = E_{3/2}$$

On the other hand, the electric dipole moments $p_x = ex$ and $p_y = ey$ belong to the irreducible representation E, so that the electric dipole σ ($E \parallel x$ or $E \parallel y$) polarized transitions are related to the following direct products:

$$E_{1/2} \times E = E_{1/2} + E_{3/2} \quad \text{and} \quad E_{3/2} \times E = 2E_{1/2}$$

Therefore, we can establish the electric dipole selection rules given in Table 7.8, and then make a proper assignment for the emission peaks of Sm^{3+} ions observed in the experimental emission spectrum of Figure 7.9.

This emission spectrum consists of five main peaks. The highest energy line peaks at 15 480 cm^{-1} and is observed under both σ and π polarization. Thus, according to the selection rules given in Table 7.8, this emission must be related to a $E_{1/2}$ ($^4G_{5/2}$) $\rightarrow$ $E_{1/2}$ ($^6H_{9/2}$) transition (otherwise, it would not display this double polarization character). This indicates that the lowest energy level of the ground state (at 0 cm^{-1}) and the emitting level (at 15 480 cm^{-1}) are both labeled by $E_{1/2}$ irreducible representations. In a similar way, the other doubly (σ and π) polarized emission lines, peaking at 15 446 cm^{-1} 15 361 cm^{-1}, are both related to $E_{1/2}$ ($^4G_{5/2}$) $\rightarrow$ $E_{1/2}$ ($^6H_{9/2}$) transitions. On the other hand, the two emission lines at 15 418 cm^{-1} and 15 297 cm^{-1} are only relevant under σ polarization (although weak emission signals are still observed under π polarization). Thus, they must correspond to transitions terminating in $E_{3/2}$ ($^6H_{9/2}$) levels.

Table 7.8   *ED selection rules in $\overline{D}_3$ symmetry for a Kramers ion*

	$E_{1/2}$	$E_{3/2}$
$E_{1/2}$	σ, π	σ
$E_{3/2}$	σ	π

Now we can construct a simplified energy-level scheme, the one shown in Figure 7.10, to account for the observed emission spectrum. It should be noted

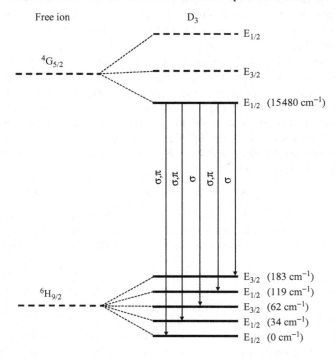

Figure 7.10   An energy-level scheme to account for the emission spectrum of Sm^{3+} ions in YAB.

that we have been able to determine the energies and label (using the irreducible representations $E_{1/2}$ and $E_{3/2}$) the energy levels of the ground state $^6H_{9/2}$ and the lowest energy level (the emitting level) of the excited state $^4G_{5/2}$. However, the other two components of the excited state (those marked with dotted lines) cannot be neither determined nor labeled from the emission spectrum given in Figure 7.9, as these levels do not participate in this de-excitation process. In any case, these levels could be located and properly labeled from polarized absorption spectra.

EXERCISES

7.1 (a) Using the character table of point symmetry group D_3 (Table 7.6), decompose the reducible representations Γ_a and Γ_b given below in irreducible representations of the D_3 group. (b) Deduce the degeneracy of the energy levels labeled by the A_1, A_2, and E irreducible representations of this symmetry group. (c) On the basis of Table 7.6, what is the number of symmetry elements of the D_3 group?

	E	$2C_3$	$3C_2'$
Γ_a	2	2	0
Γ_b	5	-1	-1

7.2 In order to develop a phosphor emitting at about 285 nm, a crystal doped with Tm^{3+} ions is grown. (a) Using the Dieke's diagram in Figure 6.1, determine the more suitable $^{2S+1}L_J$ initial and final states to produce this luminescence. (b) If the local symmetry of Tm^{3+} ions is O, construct an energy-level diagram indicating the excited and terminal energy-level components of the $^{2S+1}L_J$ states determined in (a). Use the character table of the O group (Table 7.4).

7.3 (a) For Exercise 7.2, sketch the emission spectrum that you expect (displaying only the intensity of each emission line versus an ideal unknown wavelength), considering only the degeneracy of the terminal energy levels. (b) Now establish the electric dipole selection rules and then sketch the emission spectrum that you would expect.

7.4 (a) Deduce the number and label the split components of the $^3P_0 \rightarrow {}^3H_4$ emission spectrum of Pr^{3+} ions in a crystal producing a local symmetry O around these ions (use the character table given in Table 7.4). (b) Determine the electric dipole selection rules between the crystal-field levels of the previous transition and so infer the number of peaks that you expect to appear in the emission spectrum. (c) If an axial pressure is applied to this crystal, in such a way that the new local symmetry for Pr^{3+} ions is D_4 (see the character table, in Table 7.4), sketch the new emission spectrum that you would expect according to symmetry reduction. Is it polarized?

7.5 The low-temperature absorption spectrum of Nd^{3+} ions in $LiNbO_3$ crystals corresponding to the $^4I_{9/2} \rightarrow {}^4F_{3/2}$ transition consists of two peaks at 11 253 cm^{-1} and 11 416 cm^{-1}. These two peaks are related to the split components of the $^4F_{3/2}$ excited state. This excited state splits due to the crystalline environment of the Nd^{3+} ions (C_3 symmetry). The high-energy absorption peak (11 416 cm^{-1})

Table 7.9 The character table for the $\bar{C}_3$ group. In this table, $\omega = e^{i\pi/3}$

$\bar{C}_3$	E	R	C_3	C_3R	C_3^2	C_3^2R	
A	1	1	1	1	1	1	z
$E\{$	1	1	ω^2	ω^2	$-\omega$	$-\omega$	$\}x, y$
	1	1	$-\omega$	$-\omega$	ω^2	ω^2	
$E_{1/2}\{$	1	-1	ω	$-\omega$	ω^2	$-\omega^2$	$\}$
	1	-1	$-\omega^2$	ω^2	$-\omega$	ω	
$B_{1/2}$	1	-1	-1	1	1	-1	

appears under both σ and π polarization, while the low-energy absorption peak ($11\ 253\ \text{cm}^{-1}$) is only observed for σ polarization. Using the character table of the $\overline{C}_3$ double group (Table 7.8, given below) and considering electric dipole transitions, display an energy-level diagram and label by irreducible representations the initial and terminal Stark energy levels responsible for each of these two absorption peaks.

REFERENCES AND FURTHER READING

Cavalli, E., Speghini, A., Bettinelli, M., Rámirez, M. O., Romero, J. J., Bausá, L. E., and García Solé, J., *J. Lumin.*, **102**, 216 (2003).

Duffy, J. A., *Bonding, Energy Levels and Bands in Inorganic Solids*, Longman Scientific & Technical, Haslon (1990).

Henderson, B., and Imbusch, G. F., *Optical Spectroscopy of Inorganic Solids*, Oxford Science Publications, Oxford (1989).

Jaque, D., Romero, J. J., Ramirez, M. O., Sanz García, J. A., De las Heras, C., Bausá, L. E., and García Solé, J., *Rad. Effects Defects Solids*, **158**, 231 (2003).

Tsukerblat, B. S., *Group Theory in Chemistry and Spectroscopy*, Academic Press, London (1994).

Appendix A1   The Joint Density of States

In order to obtain the frequency dependence for the joint density of states $\rho(\omega)$ (Equation (4.32)), we assume the parabolic band structure given in Figure 4.8(a). For simplicity, we suppose that the bottom of the conduction band ($E_f = E_g$) and the top of the valence band ($E_i = 0$) are both at $\vec{k} = 0$, as shown in Figure A1.1. Then, the E–k relationships are given by:

$$E_f = E_g + \frac{\hbar^2 k^2}{2m_e^*} \tag{A1.1}$$

$$E_i = -\frac{\hbar^2 k^2}{2m_h^*} \tag{A1.2}$$

where m_e^* and m_h^* are the effective masses of the electron and hole, respectively. These formulas indicate iso-energetic surfaces in k-space, as the energy does not depend on the direction of $\vec{k}$ ($E = E(|\vec{k}|)$).

Let us suppose an incident photon of energy $\hbar\omega$. The number of energy states in the frequency range $\omega \rightarrow \omega + d\omega$ (see Figure A1.1) is given by $\rho(\omega)d\omega$. This number of energy states can also be expressed as a function of the density of states in k-space, so that we can write

$$\rho(\omega)d\omega = \rho_k \Delta k \tag{A1.3}$$

where ρ_k is the number of states per unit k-volume and $\Delta k = 4\pi k_1^2 \, dk$ is the incremental volume between two spheres of radius k_1 and k_2 ($dk = k_2 - k_1$). Taking into account expressions (A1.1) and (A1.2) and Figure A1.1, these two k values can easily

An Introduction to the Optical Spectroscopy of Inorganic Solids   J. García Solé, L. E. Bausá, and D. Jaque
© 2005 John Wiley & Sons, Ltd   ISBNs: 0-470-86885-6 (HB); 0-470-86886-4 (PB)

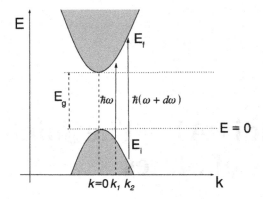

Figure A1.1 $E-k$ curves for a parabolic band structure, showing the frequency range $\omega \to \omega + \mathrm{d}\omega$ between two k values, k_1 and k_2.

be related to the photon frequencies ω and $\omega + \mathrm{d}\omega$:

$$\hbar\omega = \hbar\omega_g + \frac{\hbar^2 k_1^2}{2}\left(\frac{1}{m_e^*} + \frac{1}{m_h^*}\right) \tag{A1.4}$$

$$\hbar(\omega + \mathrm{d}\omega) = \hbar\omega_g + \frac{\hbar^2 k_2^2}{2}\left(\frac{1}{m_e^*} + \frac{1}{m_h^*}\right) \tag{A1.5}$$

Solving for k_1 and k_2, we obtain:

$$k_1^2 = \frac{2\mu(\omega - \omega_g)}{\hbar} \tag{A1.6}$$

$$k_2^2 = \frac{2\mu(\omega + \mathrm{d}\omega - \omega_g)}{\hbar} \tag{A1.7}$$

where $\mu = m_e^* m_h^*/(m_e^* + m_h^*)$ is the reduced effective mass of the electron–hole system. From expressions (A1.6) and (A1.7), we deduce that

$$k_2^2 = k_1^2 + \frac{2\mu}{\hbar}\mathrm{d}\omega \tag{A1.8}$$

On the other hand, we can also write

$$k_2^2 = (k_1 + \mathrm{d}k)^2 = k_1^2 + 2k_1\,\mathrm{d}k + (\mathrm{d}k)^2 \cong k_1^2 + 2k_1\mathrm{d}k \tag{A1.9}$$

as $(dk)^2$ is a very small quantity in comparison to the other terms. Now, combining expressions (A1.8) and (A1.9), we can obtain $(2\mu/\hbar)\,d\omega \cong 2k_1\,dk$, or

$$dk \cong \frac{\mu}{\hbar k_1}d\omega \tag{A1.10}$$

and, consequently,

$$\Delta k \cong \frac{4\pi k_1^2 \mu}{\hbar k_1}d\omega = \frac{4\pi k_1 \mu}{\hbar}d\omega \tag{A1.11}$$

Let us now determine ρ_k. This value is given by

$$\rho_k = 2 \times \frac{1}{8\pi^3} = \frac{1}{4\pi^3} \tag{A1.12}$$

where the factor of 2 is due to the fact that there are two electron spin states for each allowed k-state and the factor $1/8\pi^3$ is the density of states in k-space.[1]

Now, inserting Equations (A1.11) and (A1.12) in expression (A1.3) and using (A1.6), we obtain

$$\rho(\omega)d\omega = \frac{1}{4\pi^3} \times \frac{4\pi\mu}{\hbar}\left[\frac{2\mu(\omega - \omega_g)}{\hbar}\right]^{1/2}d\omega \tag{A1.13}$$

and, after simplifying,

$$\rho(\omega) = \frac{1}{2\pi^2}\left(\frac{2\mu}{\hbar}\right)^{3/2}(\omega - \omega_g)^{1/2} \tag{A1.14}$$

which is just the expression for the joint density of states given by Equation (4.32).

---

[1] $k = (2\pi/L)(n_x, n_y, n_z)$, where (n_x, n_y, n_z) are integers and L is a macroscopic length. Thus, it can be seen that each allowed k state occupies a k-space volume of $(2\pi/L)^3$, so that the number of states in a unit volume of k-space is $(L/2\pi)^3$. Consequently, a unit volume of material will have $(1/2\pi)^3 = 1/8\pi^3$ states per unit volume of k-space.

Appendix A2   The Effect of an Octahedral Field on a d^1 Valence Electron

We assume that the ligand ions B in Figure 5.1 are point charges located at a distance a from the d^1 central ion and that a strong crystalline field, $H_{SO} \ll H_{CF}$, is acting on that central ion. Thus, the orbital eigenfunctions ψ_{n,l,m_l} (n, l, and m_l being quantum numbers) of a single electron in a central field can be written as a product of two factors:

$$\psi_{n,l,m_l} = R_{n,l} \times Y_l^{m_l} \tag{A2.1}$$

where R_{n,m_l} is the radial part and $Y_l^{m_l}$ is the angular part.

The R_{n,m_l} functions are related to the average probability of finding an electron in an specific orbital at a distance r from the nucleus of the central ion. We do not consider this part of the function in our calculation, because it is unaffected by the crystalline field (it does not lead to energy splitting).

The spherical harmonics $Y_l^{m_l}$ describe the directional properties and can be written as follows:

$$Y_l^{m_l} = \Theta_l^{m_l}(\theta) \times (2\pi)^{-1/2} e^{im_l \phi} \tag{A2.2}$$

where the $\Theta_l^{m_l}(\theta)$ are functions of $\sin \theta$ and $\cos \theta$ (listed in S. Sugano et al., 1970) associated with the normalized Legendre polynomials, θ and ϕ being the polar and azimuthal coordinates, respectively (see Figure A2.1).

For a d^1 electron, $l = 2$ and $m_l = \pm 2, \pm 1, 0$, and the eigenfunctions $\psi_{n,2,m_l}$ can be represented in a simplified notation as the functions (m_l). The problem is

An Introduction to the Optical Spectroscopy of Inorganic Solids   J. García Solé, L. E. Bausá, and D. Jaque
© 2005 John Wiley & Sons, Ltd   ISBNs: 0-470-86885-6 (HB); 0-470-86886-4 (PB)

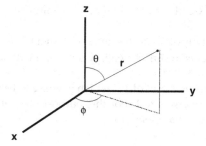

Figure A2.1   The spherical coordinates of the valence electron.

to calculate the matrix elements $H_{m_l,m_l'} = \langle m_l \,|H_{CF}|\, m_l' \rangle$ and then solve the secular equation

$$\begin{vmatrix} H_{2,2} - E & H_{2,1} & H_{2,0} & H_{2,-1} & H_{2,-2} \\ H_{1,2} & H_{1,1} - E & H_{1,0} & H_{1,-1} & H_{1,-2} \\ H_{0,2} & H_{0,1} & H_{0,0} - E & H_{0,-1} & H_{0,-2} \\ H_{-1,2} & H_{-1,1} & H_{-1,0} & H_{-1,-1} - E & H_{-1,-2} \\ H_{-2,2} & H_{-2,1} & H_{-2,0} & H_{-2,-1} & H_{2,-2} - E \end{vmatrix} = 0 \qquad \text{(A2.3)}$$

where $H_{CF} = e \times V(r, \theta, \phi)$ is the crystal-field Hamiltonian, (r, θ, ϕ) being the spherical coordinates of the d^1 electron (see Figure A2.1).

For our octahedral environment (see Figure 5.1), it can be shown (Figgis, 1961) that the electric potential created by the ligand B ions at the d^1 electron can be written as follows:

$$V(r, \theta, \phi) = 6\frac{Ze}{a} + \left(\frac{49}{18}\right)^{1/2} (2\pi)^{1/2} \left(Ze\frac{r^4}{a^5}\right)\left(Y_4^0 + \left(\frac{5}{14}\right)^{1/2} (Y_4^4 + Y_4^{-4})\right)$$

$$\text{(A2.4)}$$

where Ze is the charge of the ligand ions. Thus, we can express $V(r, \theta, \phi)$ in terms of two functions, V_1 and V_2, as:

$$V = V_1 + V_2 \qquad \text{(A2.5)}$$

where V_1 and V_2 are the first and the second terms on the right-hand side of Equation (A2.4). The matrix elements related to V_1 and V_2 can be obtained separately.

Let us first calculate the matrix element related to V_1:

$$\int_{vol} (m_l)^* \left(\frac{6Ze}{a}\right) (m_l')\, d\tau = \frac{6Ze}{a} \int (m_l)^*(m_l')\, d\tau = \begin{cases} 0 & \text{if } m_l \neq m_l' \\ \dfrac{6Ze}{a} & \text{if } m_l = m_l' \end{cases} \qquad \text{(A2.6)}$$

where the integral is extended over an infinite volume τ and the asterisk (*) refers to the conjugate complex function.

Thus, the contribution of V_1 to the matrix elements H_{m_l, m'_l} is equal to $6Ze^2/a$. Consequently, its effect is to shift the energy level of the d^1 states by this amount, but not to split this level.

Let us now consider the effect of V_2, which is going to be the term responsible for the splitting of the d^1 energy level. For this purpose, we take into account the next general property of the radial functions:

$$\int_0^\infty (R_{n,l})^* r^s (R_{n,l}) r^2 \, dr = \langle r^s_{n,l} \rangle \tag{A2.7}$$

where $\langle r^s_{n,l} \rangle$ is the average value of $r^s_{n,l}$. For d functions, $(m_l) = R_{n,2} \times Y_2^{m_l}$ and so, taking Equation (A2.7) into account, we can write:

$$\int_{vol} (m_l)^* V_2(m'_l) \, d\tau = \left(\frac{49}{18}\right)^{1/2} (2\pi)^{1/2} \langle r^4_{n,2} \rangle \left(\frac{Ze}{a^5}\right) \int_0^\pi \int_0^{2\pi} (Y_2^{m_l*} Y_4^0 Y_2^{m'_l} \sin\theta \, d\theta \, d\phi$$

$$+ \left(\frac{5}{14}\right)^{1/2} \left(Y_2^{m_l*} Y_4^4 Y_2^{m'_l*} \sin\theta \, d\theta \, d\phi + Y_2^{m_l*} Y_4^{-4} Y_2^{m'_l} \sin\theta \, d\theta \, d\phi\right) \tag{A2.8}$$

Now, taking into account the property of the spherical harmonics $Y_l^{m_l}$:

$$\int_0^{2\pi} Y_{l_1}^{m_{l_1}} \times Y_{l_1}^{m_{l_2}} \times Y_{l_1}^{m_{l_3}} \, d\phi \neq 0 \qquad \text{only for} \quad m_{l_1} + m_{l_2} + m_{l_3} = 0 \tag{A2.9}$$

and also Equation (A2.2), is easy to obtain that

$$\int (m_l)^* V_2(m_l) \, d\tau = \left(\frac{49}{18}\right)^{1/2} \times \left(Ze\frac{\langle r^4_2 \rangle}{a^5}\right) \int_0^\pi (\Theta_2^{m_l})^* \times \Theta_4^0 \times \Theta_2^{m_l} \sin\theta \, d\theta$$

$$\tag{A2.10}$$

where $\langle r^4_2 \rangle$ (the sub-index n is suppressed for simplicity) is the average value of the fourth power of the radial coordinate for the d^1 electron of the central ion. The integrals $\int_0^\pi (\Theta_2^{m_l})^* \times \Theta_4^0 \times \Theta_2^{m_l} \sin\theta \, d\theta$ are tabulated (although they can be obtained directly), so that the matrix elements $\langle m_1 | e V_2 | m_1' \rangle$ are as follows:

$$e \int (0)^* V_2(0) \, d\tau = e \left(Z e \frac{\langle r_2^4 \rangle}{a^5} \right)$$

$$e \int (\pm 1)^* V_2(\pm 1) \, d\tau = -e \frac{2}{3} \left(Z e \frac{\langle r_2^4 \rangle}{a^5} \right)$$

$$e \int (\pm 2)^* V_2(\pm 2) \, d\tau = e \frac{1}{6} \left(Z e \frac{\langle r_2^4 \rangle}{a^5} \right)$$

$$e \int (\pm 2)^* V_2(\mp 2) \, d\tau = e \frac{5}{6} \left(Z e \frac{\langle r_2^4 \rangle}{a^5} \right) \tag{A2.11}$$

Now, for the sake of simplicity, we define a parameter Dq (in CGS units), as follows:

$$Dq = \left(\frac{1}{6} \right) \left(Z \frac{e^2 \langle r_2^4 \rangle}{a^5} \right) \tag{A2.12}$$

where the factor $D = 35 Z e^2 / 4 a^5$ depends on the surrounding point charges (ligands) and the factor $q = (2/105) \langle r_2^4 \rangle$ reflects the properties of the central ion.

Now, we can construct the secular equation as follows:

(2)	(1)	(0)	(−1)	(−2)	
$Dq - E$	0	0	0	$5Dq$	
0	$-4Dq - E$	0	0	0	
0	0	$-6Dq - E$	0	0	$= 0$ (A2.13)
0	0	0	$-4Dq - E$	0	
$5Dq$	0	0	0	$Dq - E$	

This determinant is easily reduced to give:

- $E = -4Dq$ for the (1) and (−1) states;

- $E = 6Dq$ for the (0) states;

- $\begin{vmatrix} Dq - E & 5Dq \\ 5Dq & Dq - E \end{vmatrix}$ for the states (+2) and (−2), whose solutions are $E = -4Dq$ and $E = 6Dq$.

This means that the fifthly degenerate d^1 energy level splits into two levels in an octahedral crystalline field: one triply degenerate and the other doubly degenerate.

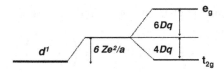

Figure A2.2   The effect of an octahedral crystalline field on a d^1 energy level.

These states are called t_{2g} and e_g, respectively, according to the usual notation based on the irreducible representations associated with these levels (see Chapter 7).

Figure A2.2 shows the effect that produces an octahedral crystalline field at the fifthly degenerate d^1 energy level. Due to the term V_1, the level is shifted to a higher energy, by an amount of energy $6Ze^2/a$. Moreover, this level splits into two levels: the t_{2g} level (triply degenerate) at an energy of $E = -4Dq$ and the e_g level (doubly degenerate) at an energy of $E = 6Dq$, both energies being measured from the shifted level. Therefore, the energy separation between these two states is $10Dq$. This splitting is manifested by the appearance of an absorption band related to the $t_{2g} \rightarrow e_g$ transition (see, for instance, the absorption spectrum of Ti^{3+} in Al_2O_3, in Figure 5.4).

REFERENCES AND FURTHER READING

Figgis, B. N., *Introduction to the Ligand Fields*, Interscience, New York (1961).
Sugano, S., Tanabe, Y., and Kamimura, H., *Multiplets of Transition Metal Ions in Crystals*, Academic Press, New York (1970).

Appendix A3   The Calculation of the Probability of Spontaneous Emission by Means of Einstein's Thermodynamic Treatment

First, we rewrite expression (5.14), which is valid for plane monochromatic waves, in a more general way, valid for a more general wavefront, as follows:

$$P_{if} = \frac{\pi}{3n^2\varepsilon_0\,\hbar^2}\rho\,\left|\vec{\mu}_{if}\right|^2\delta(\Delta\omega) \tag{A3.1}$$

where $\rho = n^2\varepsilon_0 E_0^2/2$ is the energy density of the incident electromagnetic wave. This expression leads to Equation (5.14), considering that $I = c_0\rho/n$ for plane electromagnetic waves.

Let us now assume that our two-level system is placed in a blackbody cavity whose walls are kept at a constant temperature T. Once thermal equilibrium has been reached, we can consider that our system is immersed in a thermal cavity where an electromagnetic energy density has been established. The spectral distribution ρ_ω of this energy density is given by Planck's formula:

$$\rho_\omega = \frac{\hbar\omega^3 n^3}{c_0^3\pi^2}\frac{1}{e^{\,\hbar\omega/kT}-1} \tag{A3.2}$$

which is just Equation (2.2), where $\rho_\omega d\omega$ represents the energy density in the frequency range ω, $\omega + d\omega$.

An Introduction to the Optical Spectroscopy of Inorganic Solids   J. García Solé, L. E. Bausá, and D. Jaque
© 2005 John Wiley & Sons, Ltd   ISBNs: 0-470-86885-6 (HB); 0-470-86886-4 (PB)

Since our system is in equilibrium, the number of absorption transitions $i \to f$ per unit time must be equal to the number of emission transitions $f \to i$ per unit time. Considering that the light–matter interaction processes described in Chapter 2 (Figure 2.5) are taking place, in equilibrium the rate of absorption must be equal to the rate of (stimulated and spontaneous) emission. That is:

$$B_{if} \rho_{\omega_0} N_i = B_{fi} \rho_{\omega_0} N_f + A N_f \tag{A3.3}$$

where ω_0 corresponds to the transition frequency of the system, N_i and N_f are the equilibrium populations of the initial and final levels, respectively, and B_{if} and B_{fi} are constant coefficients (the so-called Einstein B coefficients), while A is the probability of spontaneous emission (the so-called Einstein A coefficient). Thus, the probabilities of absorption and stimulated emission are written as follows:

$$P_{if} = B_{if} \rho_{\omega_0} \tag{A3.4}$$

$$P_{fi} = B_{fi} \rho_{\omega_0} \tag{A3.5}$$

Now taking into account the Boltzmann population distribution, $N_f/N_i = e^{-\hbar\omega_0/kT}$, we can write, from Equation (A3.3),

$$\rho_{\omega_0} = \frac{A}{B_{if}\, e^{\,\hbar\omega_0/kT} - B_{fi}} \tag{A3.6}$$

This expression can be compared with Planck's formula – Equations (A3.2) or (2.2) – to obtain the following two relations among the Einstein coefficients:

$$B_{if} = B_{fi} = B \tag{A3.7}$$

$$\frac{A}{B} = \frac{\hbar\omega_0^3 n^3}{\pi^2 c_0^3} \tag{A3.8}$$

Equation (A3.7) shows the equality between the probabilities of absorption and stimulated emission that we have already established for monochromatic radiation in Equation (5.15). Equation (A3.8) gives the ratio of the spontaneous to the induced transition probability. It allows us to calculate the probability A of spontaneous emission once the Einstein B coefficient is known.

At this point, we consider Equation (A3.1), which is only valid for pure monochromatic incident radiation. As we are dealing with blackbody radiation, we simulate the elemental density of radiation $\rho_\omega d\omega$ by monochromatic radiation that has the same power. According to Equation (A3.1), the corresponding probability of elemental transition (absorption or stimulated emission) dP is as follows:

$$dP = \frac{\pi}{3n^2\varepsilon_0 \hbar^2} |\mu|^2 \, \rho_\omega \delta(\Delta\omega) d\omega \tag{A3.9}$$

After integrating this equation, we obtain

$$P = \frac{\pi}{3n^2\varepsilon_0\,\hbar^2}|\mu|^2\,\rho_{\omega_0} \tag{A3.10}$$

Now, comparing this formula with Equation (A3.4) or Equation (A3.5) and taking into account the equality (A3.7), we obtain

$$B = \frac{\pi\,|\mu|^2}{3n^2\varepsilon_0\,\hbar^2} \tag{A3.11}$$

Finally, using the relationship between the Einstein A and B coefficients (A3.8) together with the previous expression, we obtain the following expression for the probability of spontaneous emission:

$$A = \frac{n\omega_0^3}{3\pi\hbar\varepsilon_0 c_0^3}|\mu|^2 \tag{A3.12}$$

Appendix A4
The Determination of
Smakula's Formula

We first redefine the cross section σ for a given transition as follows:

$$\sigma = \frac{P}{I_p} \qquad (A4.1)$$

where P is the transition rate (or transition probability) and $I_p = I/\hbar\omega$ is the photon flux of the incident wave (i.e., the intensity I of the beam divided by the energy of the photons).[1] Now, using Equations (A4.1) and (5.14) for a two-level absorption system, we can write the following expression for the cross section:

$$\sigma(\omega) = \frac{\pi}{3n\varepsilon_0 c_0 \hbar} |\mu|^2 \omega_0 g(\Delta\omega) \qquad (A4.2)$$

where ω_0 is the transition frequency and we have considered a certain line shape $g(\Delta\omega)$ for the transition probability instead of a Dirac δ-function. We thus realize how the cross section is related to the parameters of the material, such as $|\mu|^2$ and $g(\Delta\omega)$, and to the frequency of the incident light, ω_0.

---

[1] This definition is consistent with that given in Chapter 1 (Equations (1.5) and (1.6)) in relation to the absorption coefficient. To show this, we consider a two-level system as the absorbing medium and assume that the illumination intensity (or photon flux) is low. Therefore, we can assume that all of the absorbing atoms (of density N) are in the ground state. Thus, the decrease in the photon flux in a traversed thickness dz is given by $dI_p = -PN dz = -I_p \sigma N \, dz$. Integration of this equation leads to $I = I_0 e^{-\sigma N z}$ (where we have again used intensity units instead of photon flux units). This is just the Lambert–Beer law (Equation (1.4), provided that $\alpha = \sigma N$, as defined in Equation (1.6).

An Introduction to the Optical Spectroscopy of Inorganic Solids   J. García Solé, L. E. Bausá, and D. Jaque
© 2005 John Wiley & Sons, Ltd   ISBNs: 0-470-86885-6 (HB); 0-470-86886-4 (PB)

Using Equations (5.17) and (A4.2), the cross section can be written in terms of the spontaneous emission probability:

$$\sigma(\omega) = \frac{\pi^2 c_0^2}{n^2 \omega_0^2} A g(\Delta\omega) \tag{A4.3}$$

We can integrate over the frequency range corresponding to the absorption band and so write

$$\int \sigma(\omega)\,d\omega = \frac{\pi^2 c_0^2}{n^2 \omega_0^2} A \int g(\Delta\omega)\,d\omega = \frac{\pi^2 c_0^2}{n^2 \omega_0^2} A \tag{A4.4}$$

Using expression (1.6), we can write a similar expression but for the absorption coefficient:

$$\int \alpha(\omega)\,d\omega = N \frac{\pi^2 c_0^2}{n^2 \omega_0^2} A \tag{A4.5}$$

Now, inserting expression (5.20) for A into equation (A4.5), we obtain

$$\int \alpha(\omega)\,d\omega = \frac{1}{4\pi\varepsilon_0} \frac{2\pi^2 e^2}{mc_0} \left[\left(\frac{E_{loc}}{E}\right)^2 \frac{1}{n} \right] \times fN \tag{A4.6}$$

This expression is exactly coincident with Equation (5.21), which leads to Smakula's formula, Equation (5.22), after inserting numerical values and the local field correction factor for centers of high symmetry.

Index